International Max Planck Research School (IMPRS)
for Maritime Affairs
at the University of Hamburg

Hamburg Studies on Maritime Affairs
Volume 10

Edited by

Jürgen Basedow · Ulrich Magnus
(Editors)

Pollution of the Sea – Prevention and Compensation

Springer

Professor Dr.Dr.h.c. Jürgen Basedow, LL.M. (Harvard)
Max Planck Institute for Comparative
and International Private Law
Mittelweg 187
20148 Hamburg
Germany
basedow@mpipriv.de

Professor Dr. Ulrich Magnus
Fachbereich Rechtswissenschaft
Universität Hamburg
Schlüterstraße 28
20146 Hamburg
Germany
ulrich.magnus@jura.uni-hamburg.de

Library of Congress Control Number: 2007929778

ISSN 1614-2462

ISBN 978-3-540-73395-9 Springer Berlin Heidelberg New York

Springer is a part of Springer Science+Business Media

springer.com

Production: LE-TEX Jelonek, Schmidt & Vöckler GbR, Leipzig
Cover-design: WMX Design GmbH, Heidelberg

SPIN 12084784 64/3180YL - 5 4 3 2 1 0 Printed on acid-free paper

Preface

In May 2005, the International Max Planck Research School for Maritime Affairs organized an international conference addressing marine pollution. At the conference venue, the International Tribunal for the Law of the Sea, academics and practitioners debated recent developments from an interdisciplinary perspective. The papers shed light upon ecological aspects, upon the efforts of the international community to prevent and combat marine pollution by appropriate institutions and regulations, and upon the compensation of losses which have occurred despite those efforts. The present volume collects the papers of the conference and at the same time conveys an impression of the diversity of subjects discussed in the International Max Planck Research School for Maritime Affairs.

The editors of this book have been supported by advice and help from several persons: In particular we are indebted to Dr. Silke Knaut and Dr. Anatol Dutta for their editorial assistance, to Anna-Maria Hubert who, as a native speaker, carried out the linguistic revision of several papers, and to Ingeborg Stahl for the preparation of the manuscript.

Hamburg, April 2007 *Jürgen Basedow/Ulrich Magnus*

Contents

Contributors IX

Welcome Address and Introduction
Marine Pollution as a Topic of Research and Policy
Jürgen Basedow 1

Part I: Ecological Aspects of Marine Pollution 5

Survey: Sources, Paths and Effects of Marine Pollution
Jürgen Sündermann 7

Modelling the Fate of Persistent Toxic Substances in the North Sea: γ-HCH and PCB 153 Multiyear Simulations
Tatjana Ilyina 15

Rebuilding the Eastern Baltic Cod Stock in a System of Change – An MPA Approach
Christine Röckmann 23

Part II: Prevention of Marine Pollution – Institutional Foundations 39

Prevention of Marine Pollution: The Contribution of IMO
Thomas A. Mensah 41

The Contribution of the European Union to Marine Pollution Prevention
Ludwig Krämer 63

HELCOM's Contribution to the Prevention of Marine Pollution
Anne Christine Brusendorff 85

Transport of Hazardous and Noxious Goods by Sea – The IMDG Code
Meltem Deniz Güner 95

Part III: Compensation for Marine Pollution ... 111

Origins and Compensation of Marine Pollution – A Survey
Peter Ehlers ... 113

Maritime Pollution – Compensation or Enforcement?
Rüdiger Wolfrum ... 129

The International Oil Pollution Compensation Funds and the International Regime of Compensation for Oil Pollution Damage
Måns Jacobsson ... 137

Compensation by the Coastal States – The *Prestige* Disaster
Juan L. Pulido ... 151

International Convention on Civil Liability for Bunker Oil Pollution Damage, 2001 – Liability and Insurance Aspects
Ling Zhu ... 171

Closing Remarks
Ulrich Magnus ... 181

Contributors

Jürgen Basedow

Dr. iur., Dr. h.c., LL.M. (Harvard), is Professor at the Law Faculty of the University of Hamburg and since 1997 the director of the Max Planck Institute for Comparative and International Private Law in Hamburg. He is a member of various governmental advisory committees in the field of transport, insurance and competition law and speaker of the International Max Planck Research School for Maritime Affairs at the University of Hamburg. His key research areas are international economic law, the law of transport, shipping and insurance, private international law and comparative law.

Anne Christine Brusendorff

Has a Master of Laws from the University of Copenhagen, Denmark and London School of Economics and Political Science, United Kingdom, and a Ph.D. from the University of Copenhagen, Denmark. She has dealt with the maritime and response issues within HELCOM from 1998, and has been HELCOM Executive Secretary since 2003.

Peter Ehlers

Born in 1943, Dr. iur., President of the German Federal Maritime and Hydrographic Agency, German representative at several international organizations (IHO and IOC/UNESCO), Past Chairman of the Helsinki Commission and the International Hydrographic Conference, board member of several maritime and marine scientific institutions, Professor of law at the University of Hamburg, Director of the International Max Planck Research School for Maritime Affairs at the University of Hamburg, editor and author of numerous publications on the law of the sea and marine issues.

Meltem Deniz Güner

Studies of Law at Istanbul University Faculty of Law; LL.M. in Private Law at Istanbul University Faculty of Law; LL.M. in Admiralty Law at Tulane Law School, U.S.A; member of the Istanbul Bar; research assistant at the Maritime & Insurance Law Department of Istanbul University Faculty of Law (on leave during her studies at the IMPRS); Scholar of International Max-Planck Research School for Maritime Affairs at the University of Hamburg ("IMPRS").

Tatjana Ilyina

Born in 1978 in Sochi, Russia. She studied Hydrometeorology in the Russian State Hydrometeorological University in Saint Petersburg leading to BSc and MSc degrees obtained in 2000 and 2002 respectively. During her graduate studies she taught Introduction to Meteorology and Scientific English to undergraduate students. In 2002 she became a scholar of the International Max Planck Research School for Maritime Affairs at the University of Hamburg. Within her doctoral project she developed a numeric model for studying the behaviour of persistent organic pollutants in the marine environment. She finished her doctorate studies under supervision of Prof. Dr. Jürgen Sündermann in 2006.

Måns Jacobsson

Director of the International Oil Pollution Compensation Funds since January 1985, following a career in Sweden where he served as a judge and thereafter became Assistant Under Secretary and Head of Department for International Affairs in the Swedish Ministry of Justice. He was later appointed President of Division of the Stockholm Court of Appeal. He is also Visiting Professor of the World Maritime University in Malmö, Sweden and of Shanghai Maritime University, Shanghai, People's Republic of China.

Ludwig Krämer

Judge at the Landgericht Kiel from 1969 until 2004; worked in the Environmental Directorate General of the Commission of the European Community from 1972 until 2004. Since then, he is retired. He is lecturer at the College of Europe (Brugge), Honorary Professor at the University of Bremen and Visiting Professor at the Universities of Copenhagen, London and Gent. He specialised on EC environmental law and published extensively on this topic, among others the books *EC Environmental Law* (6th ed. 2006) and *Casebook on EU Environmental Law* (2002), furthermore some 170 articles.

Ulrich Magnus

Dr. iur., is Professor at the Faculty of Law at the University of Hamburg; Chair for civil law, private international law and comparative law; Judge (part-time) at the Court of Appeal of Hamburg; Executive Vice-director of the European Centre for Tort and Insurance Law in Vienna; Germany's National Correspondent at UNCITRAL; Co-speaker of the International Max Planck Research School for Maritime Affairs at the University of Hamburg; Member of the German Council for Private International Law, of the European Group on Tort Law and of the European Acquis Group.

Thomas A. Mensah

Judge of the International Tribunal for the Law of the Sea, from 1996 to 2005, the first President of the Tribunal from 1996 to 1999. Prior to his election to the Tribunal, Lecturer of Law University of Ghana and Dean of the Faculty of Law,

Associate Legal Officer at the International Atomic Energy Agency, Vienna and Assistant Secretary-General and Director of Legal Affairs and External Relations at the International Maritime Organization (IMO); Cleveringa Professor of Law at Leiden University and Professor of Law and Director, Law of the Sea Institute at the University of Hawaii; from 1995 to 1996 High Commissioner (Ambassador) of Ghana to the Republic of South Africa and Chairman of the F4 (Environmental Claims) Panel, United Nations Compensation Commission (UNCC), Geneva from 2000 to 2005. Since 1989 a Member of the "Institut de Droit International", Titular Member of the "Comité Maritime International" (CMI) and Member of the Advisory Council of the British Institute of International and Comparative Law.

Juan Luis Pulido Begines

Born in 1965. Bachelor of Law (1989); Doctor of Law (1995). His present post is Professor of Maritime and Commercial Law, University of Cádiz (Spain), and Director of the *Master in Port Administration*, Port of Algeciras (Spain). Writings (selection): *Los contratos de remolque marítimo*, Barcelona, 1996; *El derecho de información del socio en la sociedad de responsabilidad limitada*, Madrid, 1997; *El Derecho de información del accionista*, Madrid, 1998; *Seguro de mercancías y seguro de responsabilidad civil del porteador terrestre,* Barcelona, 2001; *Las averías y los accidentes de la navegación marítima y aérea*, Madrid, 2003; *La responsabilidad frente a terceros de las sociedades de clasificación*, Bilbao, 2006.

Jürgen Sündermann

Born in 1938 in Oppeln/Silesia. Dr. rer. nat. and habilitation in oceanography (both University of Hamburg). 1955-1962 study of mathematics and physics at Berlin (Humboldt University) and Münster. Diploma in mathematics. 1971-1978 Professor of Electronical Computing, University of Hannover. 1978 Professor of Oceanography, University of Hamburg. 1978-2003 Director, Institute of Oceanography and Centre of Marine and Climate Research, University of Hamburg. 1988 Honorary Professor, University of Qingdao/China. 1994 Member, Polish Academy of Sciences.

Rüdiger Wolfrum

Dr. iur., is Professor for national public and international public law at the Law Faculty of the Ruprecht-Karls-University of Heidelberg and since 1993 director and scientific member of the Max Planck Institute for Comparative Public Law and International Law, President of the International Tribunal for the Law of the Sea (since 2005), "Honorarprofessor" of the University of Hamburg, President of the German Society for International Law (Deutsche Gesellschaft für Völkerrecht, since 2005), Director of the Rhodes Academy Ocean Law and Policy. He has published widely in various fields of international public law, thereby focusing on the law of the sea, the law concerning Antarctica, environmental law as well as on human rights and United Nations issues.

Ling Zhu

Born in 1979 in China. She studied Law in the Department of Law, Hunan Normal University and obtained the LL.B. degree in 2001. In the same year, she went to England and started her LL.M. study in the School of Law, University of Southampton and she obtained the LL.M. degree in Maritime Law in February 2003. In 2002, she was accepted as a scholar and Ph.D. student by the International Max Planck Research School for Maritime Affairs at the University of Hamburg. She finished her doctorate study under the supervision of Professor Dr. Dr. h.c. Jürgen Basedow in 2006.

Welcome Address and Introduction

Marine Pollution as a Topic of Research and Policy

Jürgen Basedow

In April 2002, the International Max Planck Research School for Maritime Affairs at the University of Hamburg started its work. It was established through the cooperation of various disciplines including ecology, meteorology, economics, and law. The twelve scholars at this school represent many institutions: the University of Hamburg Institutes for Oceanography, for Maritime Law and the Law of the Sea, and for Foreign and International Private Law, and the Max Planck Institutes for Meteorology, for International Law, and for Comparative and International Private Law. Anyone familiar with the peculiarities of research knows that such interdisciplinary research is rather uncommon. Indeed, there does not appear to be another Max Planck Research School that involves such a wide range of disciplines. Lawyers may well consider interdisciplinary research simply to be a private lawyer and a public lawyer discussing the same subject. In recent years, interdisciplinary research in law and economics or law and social and political science has become more common; but cooperation between lawyers and scientists remains exceptional.

What has been the reason for this willingness to cooperate? It is a common observation that the increased use of maritime space for various potentially conflicting purposes requires intensified research into marine affairs. Scientists who have dealt with the ecological aspects of the increased use of the seas have deplored that their warnings of a long-term deterioration of the marine environment have had little effect in political and legal arenas. Economists have started to consider the marine environment a scarce resource, and as a consequence, economic theories on the efficient use of scarce resources are becoming relevant to the exploitation of the seas. The legal regulations pertaining to specific aspects of maritime law have been the concern of lawyers for many years. In fact, maritime law conventions focusing on single issues such as collisions at sea were already being negotiated more than 100 years ago. However, during the last twenty years a new phenomenon has aroused the interest of legal scholars, namely, the increasing number of conflicts arising from different claims to use the marine environment. It is becoming increasingly apparent that the seas are no longer solely used for traditional shipping and fishery purposes, but also as a sewer for industrial complexes and continental populations, as recreation areas for tourists, for cable communication purposes, for fish and seafood farming, as sites for wind power production, as a kind of cooling reservoir for land-based power plants, for off-shore mining and oil and gas drilling, and more. In the future, use of marine resources will intensify. The European Commission has recently stressed the need for an "all-embracing maritime policy aimed at developing a thriving maritime economy and the full potential of sea-based activity in an environmentally sustainable manner."

The increase of potentially conflicting uses gives rise to the prediction of an increase of the number of conflicts of interests and a change in their quality. Such conflicts used to be limited to parties involved in the same type of exploitation, for example, a conflict between various parties all involved in shipping. However, we can now predict a growing number of cross-exploitation conflicts that may include the following: negligent navigation that damages or destroys communication cables, the warming of the bay water used for cooling purposes of a nuclear power plant preventing its use for fish farming purposes, an oil spill on the high seas polluting beaches that are essential for tourism in the coastal zone. One consequence of a rise in cross-exploitation conflicts will be a need for a further juridification of relations in maritime spaces, which in the past were largely characterized by the absence of law and sovereignty.

It is this expectation that explains the interdisciplinary cooperation occurring at the International Max Planck Research School for Maritime Affairs. While not all of the research projects conducted within the school directly focus on the increase of conflicting uses of the marine environment, this aspect of our activities is in the very centre of this conference on "the pollution of the seas – prevention and compensation". This conference will attempt to shed some light on the various aspects of the matter treated in the Research School. Information on the ecological aspects of marine pollution will provide a basis for understanding the need for prevention. The concept of prevention will be discussed using an economic perspective and by taking into account what has already been achieved by various organizations active in this field. Further contributions will deal with situations where prevention has failed and compensation issues arise, addressing questions such as: Who shall pay? Who can be identified as the responsible person? How can claims be enforced? These and other questions give rise to a great many legal ramifications.

The speakers participating in this conference have been recruited from three groups. The first group consists of scholars from the Research School who will present some results from their dissertations. The second group includes directors of the Research School who will present their views on more comprehensive aspects. The final group is comprised of distinguished lawyers and scholars from outside the Research School who have accepted our invitation to contribute their expertise to our conference. Most of the members of the latter group represent international organizations and, thus, must cope with the problems outlined above in their everyday work. We are particularly grateful that they are willing to share their experience and expert knowledge with us.

This conference recalls and continues a long tradition of ecological research on the marine environment conducted in Hamburg dating back to the 1960s. Lawyers' interest in this topic is of a more recent origin. It has been triggered by ecological disasters and their legal aftermath, such as the adoption of various international conventions, including those provisions relating to the marine environment in the United Nations Convention on the Law of the Sea. Our conference has been convened on the premises of the International Tribunal for the Law of the Sea which is endowed with the task of enforcing those provisions. This is a particularly appropriate venue, and I am very pleased to have the former president, Thomas Mensah, and the German judge, Rüdiger Wolfrum, among the contributors. Before leaving the floor to the first speaker, I would like to express my deep gratitude to them, and to the chancellor of the International Tribunal for granting access to us to these magnificent rooms. I wish a warm welcome to you all.

Part I:

Ecological Aspects of Marine Pollution

Survey: Sources, Paths and Effects of Marine Pollution

Jürgen Sündermann

I. Introduction

In addition to natural risks such as rising sea levels, storm surges, or tsunami waves, the pollution of the marine environment represents a serious threat to coastal inhabitants. However, contrary to the first factors mentioned, pollution does not present a direct danger to human life. Societies have recently started to become aware of marine pollution, however, awareness is increasing rather slowly and is not sufficiently developed everywhere. Nevertheless, pollution of the sea damages the marine ecosystem irreversibly over long time scales, endangering a broad spectrum of resources, from seafood to recreational spaces. The struggle against marine pollution requires environmental knowledge within society, political resolve, and money. Industrial nations, which are also the biggest polluters, are meeting these criteria to some extent, but among them there is no uniform position on marine pollution. Developing countries are in danger of repeating the environmental mistakes of previous decades, but on a greater magnitude. The current state of the sea is characterized by considerable pollutant load in the shelf regions of the most industrialized nations (e.g., the Northwest European shelf), but there is also a trend of improvement in these areas. Sea shelves of less developed countries (e.g., the East Asian waters) exhibit an increasing burden. Industrial nations must share their experiences and provide conceptual and financial help to less developed countries.

Since the basic problems of marine pollution are similar in all of the world's oceans, the following examples from the North Sea can be taken as representative. The North Sea is one of the best-monitored and extensively investigated regions in the world (Sündermann, 1994).

II. The Pollution Load of the Seas

In Table 1 (World Watch Institute,<www.gdrc.org/oceans/marine-pollution.html>) the major classes of pollutants are compiled, together with their sources and effects. Oil accidents at sea are probably the most spectacular of these pollutants, as they affect the marine ecosystem and human living areas over a certain time period. Oil spills remain a permanent environmental threat; however, their remediation by various natural processes occurs readily so that nature recovers relatively quickly.

Persistent toxins which remain in the ecosystem for a long time and accumulate within marine organisms are much more dangerous than oil spills. They include heavy metals such as mercury (Hg), cadmium (Cd), zinc (Zn), copper (Cu), lead (Pb). Their input has become significantly smaller, but they stay in the system (mostly in sediments) and can be reactivated. Organic pollutants, such as PCB, HCH, DDT, are also persistent toxins. Organic pollutants have not been as thoroughly investigated, and today, represent a greater hazard than trace metals.

It may seem odd that nutrients be considered as pollutants, but their excessive supply produces algal blooms and oxygen depletion in the sea, and lead to changed marine communities over the long term. Eutrophication must be seen as a serious threat to marine ecosystems at present.

Table 1 lists further impacts on the sea, which separately considered, may appear minor; however, the combined effects of such pollutants can impair living conditions to the extent that the existence of organisms becomes impossible.

Table 1: Types, sources, and effects of marine pollution. Compiled by World Watch Institute.

Sources and Effects of Marine Pollution

Type	Primary Source/Cause	Effect
Nutrients	Runoff approximately 50% sewage, 50% from forestry, farming, and other land use. Also, airborne nitrogen oxides from power plants, cars, etc.	Promote algal blooms in coastal waters. Decomposing algae depletes water of oxygen, killing other marine life. Can spur algal blooms (red tides), releasing toxins that can kill fish and poison people.
Sediments	Erosion from mining, forestry, farming, and other land-use; coastal dredging and mining.	Cloud water; impede photosynthesis below surface waters. Clog fish gills. Smother and bury coastal ecosystems. Carry toxins and excess nutrients.
Pathogens	Sewage, livestock.	Contaminate coastal swimming areas and seafood. Cause cholera, typhoid and other diseases.
Alien Species	Several thousand per day transported in ballast water; also spread through canals linking bodies of water and fishery enhancement projects.	Out-compete native species and reduce biological diversity. Introduce new marine diseases. Associated with increased incidence of red tides and other algal blooms, a problem in major ports.

Type	Primary Source/Cause	Effect
Persistent Toxins (PCBs, heavy metals, DDT, etc.)	Industrial discharge; wastewater discharge from cities; pesticides from farms, forests, home use, etc.; seepage from landfills.	Poison or cause disease in coastal marine life, especially near major cities or industry. Contaminate seafood. Fat-soluble toxins that bio-accumulate in predators can cause disease and reproductive failure.
Oil	46% from cars, heavy machinery, industry, and other land-based sources; 32% from oil tanker operations and other shipping; 13% from accidents at sea; remaining sources include offshore oil drilling and natural seepage.	Low-level contamination can kill larvae and cause disease in marine life. Oil slicks kill marine life, especially in coastal habitats. Tar balls from coagulated oil litter beaches and coastal habitat. Oil pollution is down 60% from 1981.
Plastics	Fishing nets; cargo and cruise ships; beach litter; wastes from the plastics industry and landfills.	Discarded fishing gear continues to catch fish. Other plastic debris entangles marine life or is mistaken for food. Plastics litter beaches and coasts and may persist for 200 to 400 years.
Radioactive substances	Discarded nuclear submarine and military waste; atmospheric fallout; industrial wastes.	Create "hot spots" of radioactivity. Can enter food chain and cause disease in marine life. Accumulate in top predators and shellfish, which are eaten by people.
Thermal	Cooling water from power plants and industrial sites.	Kill off corals and other temperature-sensitive sedentary species. Displace other marine life.
Noise	Supertankers, other large vessels and machinery.	Can be heard thousands of kilometers away under water. May stress and disrupt marine life.

The Quality Status Report 2000 (OSPAR Commission 2000) for the Greater North Sea states that the input of persistent toxic substances and nutrients has been reduced with regard to "highest impact" pollutants such as trace organic contaminants and nutrients and "upper intermediate impact" pollutants such as oil and heavy metals. Nevertheless, the marine ecosystem is still very sensitive and vulnerable.

III. The Dispersal of Pollutants in the Sea

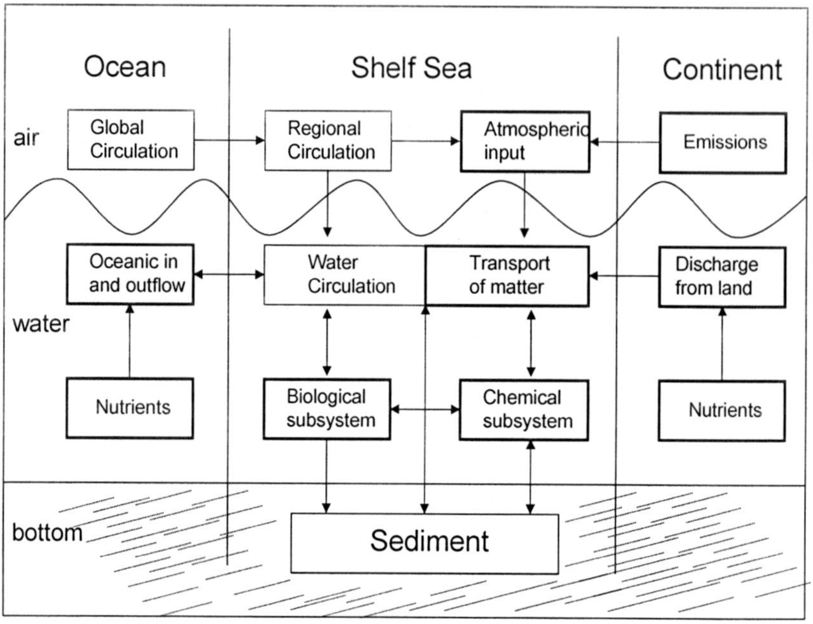

Fig. 1: Hydrodynamic interactions (thin frames) and fluxes of matter (thick frames) in a shelf sea.

Pollutant sources are mainly land based, e.g., industrial by-products, urban sewage, and nutrient release from agriculture. Transport media are flowing water (rivers) or flowing air (wind). Only some of the pollutants are directly discharged into the sea, e.g., via oil accidents or deep-sea mining. The open ocean adjacent to a shelf sea can also be a contaminant source, e.g., nutrients.

Figure 1 shows the hydrodynamic interactions and fluxes of matter in a shelf sea. The global atmospheric circulation drives the air masses above the shelf region, which in turn force, together with the incoming ocean waves, the shelf currents. These currents transport dissolved and suspended substances that enter the water body from the atmosphere, sediment, ocean, and land (external sources). Fluxes of matter can also be directed outwards, e.g., into the sediment (external sinks). Furthermore, substance concentrations can be changed within the system via chemical or biological transformations (internal sources and sinks, e.g., decay or enrichment). The direction and intensity of transport in the water are controlled by the external forces described above and by the topography of the sea. The shape of a shelf basin and the regional climate essentially determine the spatial and temporal patterns of pollutant concentrations.

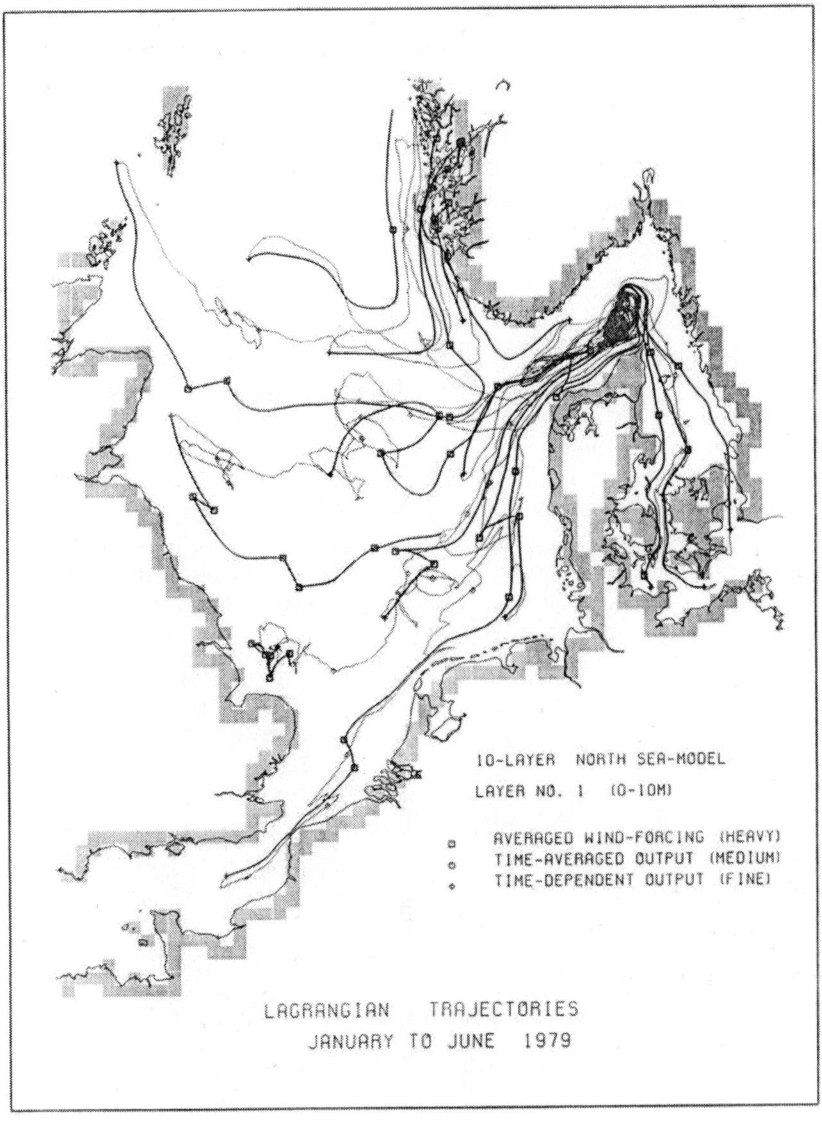

Fig. 2: Particle trajectories in the surface layer of the North Sea within 6 months for mean climatic conditions (thick lines) and the real period January to June 1979 (thin lines).

Figure 2 shows calculated trajectories of conservative, passive tracers released into the surface layer of the North Sea at different positions for half a year; both the mean climatological and the real forcing of January to June 1979 are shown (Backhaus, pers. comm.).

The basic pattern of transport from the English coast along the continental border into Scandinavian waters (anticlockwise) is present in both cases, but there are strong inter-annual and decadal variations. The stochastic nature of the atmosphere causes major concentration changes, even where there is a constant input of pollutants. This hinders any targeted control of contamination. After, periods on the order of months to years, particles leave the North Sea and enter the Atlantic Ocean, or are deposited as sediment off the Norwegian coast.

IV. Conclusions and Outlook

The ecological state of the European shelf seas, e.g., the Northwest European shelf, has been improved by reducing pollutant inputs, but this region remains vulnerable. Extended algal blooms or the death of seal populations can happen at any time, if unfavourable weather conditions or pathogens appear. Consequently, the input of pollutants must be further reduced, an achievement that will require time and continuous, sustainable environmental policy. Figure 3, depicting the effect of a 50% reduction in nutrient discharge from all North Sea rivers, shows that even drastic countermeasures only result in a moderate improvement of the system's ecological state (from Lenhart, in: Sündermann et al., 2002).

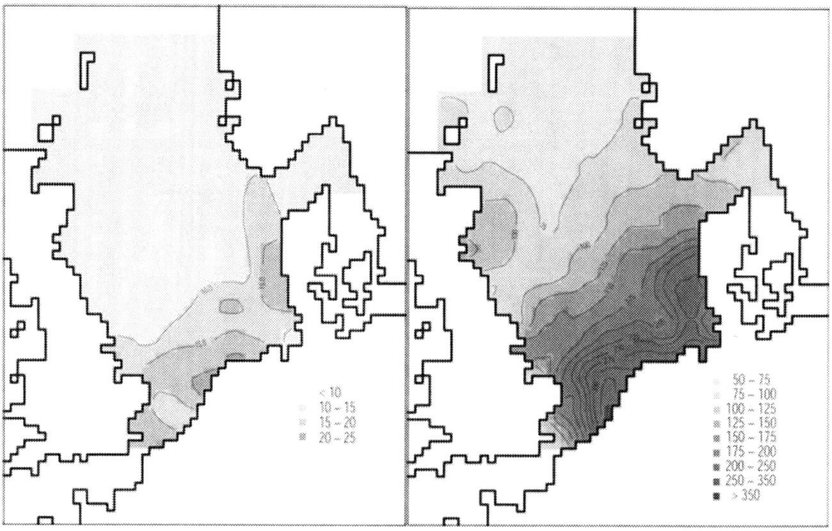

Fig. 3: Algal biomass (in grams carbon per square meter and year) in the North Sea (left). A reduction of all nutrient discharges through the rivers by half would only diminish primary production near the coast (right, numbers are % reduction).

New chemicals with unknown effects on the ecosystem are continuously entering the sea, e.g., hormones or drugs. As long as the environmental impacts of such chemicals are uncertain, the only protection strategy is to apply the precautionary principle, i.e., the release must be stopped if there is justified doubt regarding possible harm.

The climate system controls the fluxes, and thus, the effects of marine pollutants. Any change in climate results in a change in the loading state of a shelf sea, even in cases where there is a constant discharge of contaminants.

Figure 4 shows the four basic circulation patterns of the North Sea depending on wind direction (modified from Backhaus, in Sündermann et al., 2002). Prevailing winds from the southwest cause a strong anticlockwise circulation with short flushing times. As such, the current climate is favourable for the marine ecosystem. Any climate shift modifying these weather statistics may extend periods of stagnation, and hence, weaken nature's potential to heal itself.

Further research needs to be conducted regarding the prognostic modelling of the marine ecosystem on a decadal scale for different global change scenarios, e.g., adoption of measures that reduce pollution, or shifts in climate. Key issues are the understanding and quantification of processes and model validation. For this, a better database is necessary.

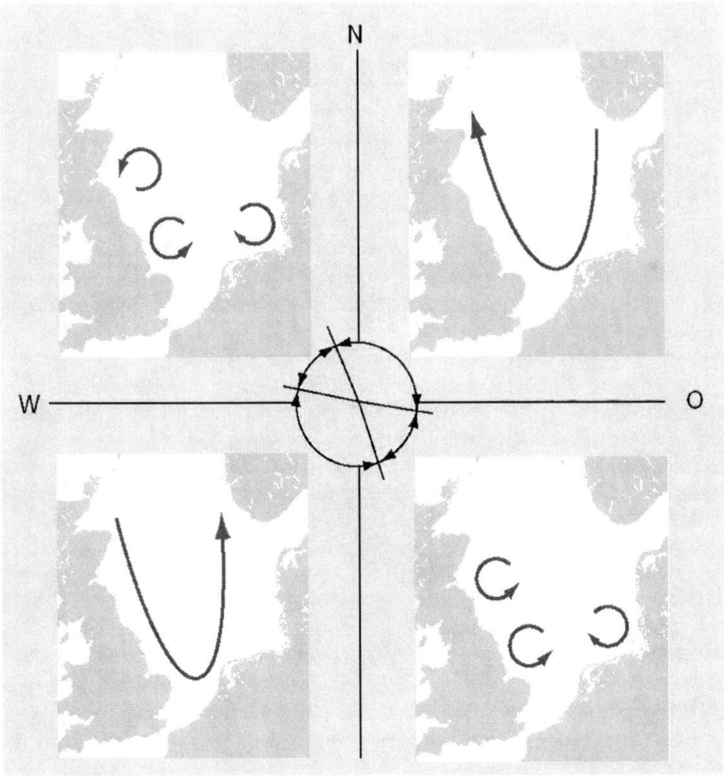

Fig. 4: Basic circulation patterns in the North Sea depending on the four sectors of prevailing winds.

References

OSPAR Commission 2000. Quality Status Report 2000. Region II – Greater North Sea, OSPAR Commission London, 136 +xiii pp.

Sündermann, J., S. Beddig, G. Radach and H. Schlünzen, 2002. The North Sea – Problems and Research Needs, Centre for Marine and Climate Research, University of Hamburg, 64 pp.

Sündermann, J. (ed.), 1994. Circulation and contaminant fluxes in the North Sea, Springer, 654 pp.

Modelling the Fate of Persistent Toxic Substances in the North Sea: γ-HCH and PCB 153 Multiyear Simulations

Tatjana Ilyina

I. Introduction

Persistent toxic substances (PTS) are organic chemicals that are environmentally persistent and harmful to human health and to the environment. Bioaccumulation or increase in concentration of a pollutant from the environment to the first organism in a food chain refers to how pollutants enter a food chain. They can be released into the environment in various ways including during their production, application (e.g., pesticides), or combustion (e.g., dioxins). Whether produced by natural or anthropogenic processes, PTS have a particular combination of physical and chemical properties allowing them to remain intact for exceptionally long periods after release into the environment. PTS migrate between different environmental compartments and undergo long-range transport (LRT) by natural processes in both the atmosphere and oceans, thus becoming ubiquitous global contaminants. PTS are distributed throughout the oceans as a consequence of atmospheric deposition and direct introduction into aquatic systems. Scientific and political interest in the fate and behaviour of PTS in the environment arises from concern over human exposure to these chemicals and their discovery in pristine environments far from source regions. There is international interest in reducing and (possibly) eliminating releases of PTS, and in reducing risks to regional and global environments. International agreements, such as the UNEP Stockholm Convention on Persistent Organic Pollutants, the UNECE Convention on Long-range Transboundary Air Pollution, and the OSPAR Convention, require assessment criteria of the environmental risks posed by PTS based on sound scientific knowledge and models.

The North Sea is a region particularly vulnerable to PTS, since highly industrialised countries releasing large amounts of PTS surround it. Monitoring campaigns show that nearly all detected PTS were found in the North Sea deriving from their persistence in seawater and LRT from distant sources. Since the beginning of the 1990s, concentrations of PTS in the North Sea have been dropping, however present-day levels still threaten the environment. It is commonly thought that oceans may be the ultimate sink for some PTS. The North Sea, a shelf and coastal sea area, is a different repository than oceans as there is a larger influx from rivers, higher atmospheric concentrations over the coastal waters, a greater proportion of sedimentation processes, and different hydrographic conditions (Sündermann, 1994). As primary pollutant sources are reduced, remobilisation from previous repositories, such as shelf and coastal seas can act as secondary sources to the atmosphere. The role of sea shelves in cycling of PTS has not been adequately researched, with only a few studies performed so far.

The objective of the present study was to advance the understanding of the fate of PTS in the aquatic environment as basis for realistic estimates of their spatial and temporal distribution and pathways in the North Sea. With such information, current and future exposure of the North Sea to contamination by PTS can be addressed.

II. Model Description and Experimental Set-up

A 3D Fate and Transport Ocean Model (FANTOM) was designed to study the long-term behaviour of PTS in the North Sea. The model focuses on quantifying the distribution of contaminants and their pathways, e.g., riverine and atmospheric inputs and inflows from water masses adjacent to the North Sea. This model accounts for major processes influencing the fate of PTS in aquatic systems, such as the transport of PTS by ocean currents, exchange between the atmosphere and sea surface via deposition (settling of particles or gases from the atmosphere on the sea surface) and volatilisation (vaporisation of a dissolved sample), degradation, redistribution between phases, and net sedimentation (sinking or re-suspension) on the seabed. Transport processes in FANTOM are driven by ocean currents calculated from the distributions of the flow field available from ocean circulation models. In this study FANTOM was coupled with Hamburg Shelf Ocean Model (HAMSOM) (Pohlmann, 1996), providing necessary hydrographic data fields for the North Sea (figure 1).

The simulation was carried out for the period from July 1995 to December 2001 with a time step of 20 min. The model covers the southern and central North Sea up to 57°N. Horizontal resolution of the model is about 3 km. Vertically, the model has 21 layers of varying depth, with 5 m thickness for the upper 50 m and 10 m for the deeper layers. Initial distributions of PTS concentrations and their concentrations on sea boundaries (i.e., the Baltic Sea, Atlantic Ocean and the English Channel) were based on non-filtered samples and therefore represent fractions of these chemicals in the dissolved phase and bound to particulates suspended in sea water. Boundary conditions at the air-sea interface are based on measured concentrations of PTS in the air and in precipitation, available through the EMEP monitoring programme. Daily fresh water discharges and monthly mean concentrations of PTS in the rivers were taken from the monitoring programmes of the corresponding environmental agencies.

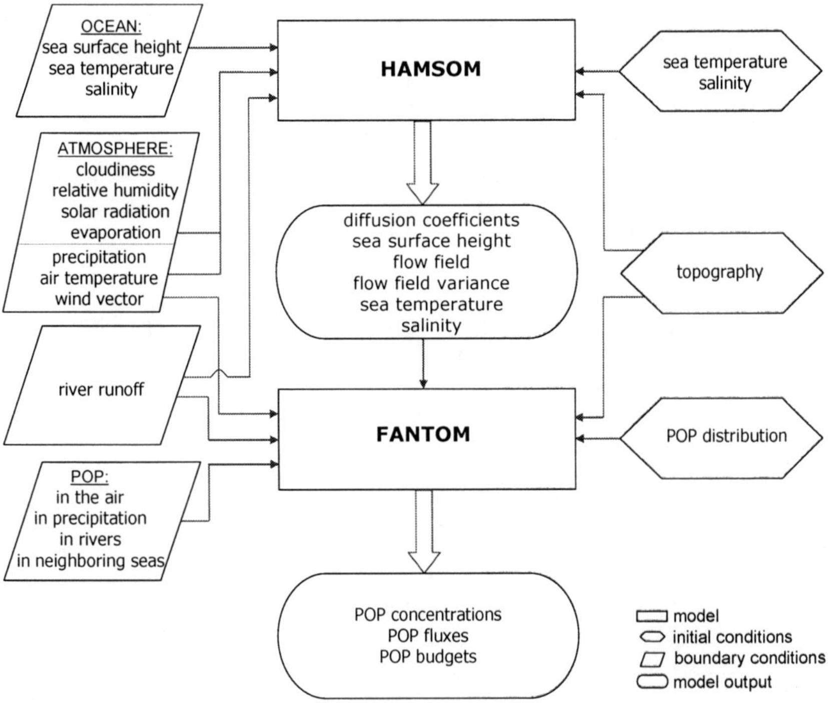

Fig. 1: The input-output diagram for the 1995-2001 integration of HAMSOM and FANTOM.

FANTOM includes necessary mechanisms to calculate environmental fate and pathways of selected PTS in the North Sea, including sources and sinks. The model design accounts for both diffuse (e.g., atmospheric deposition, resuspension or inflow through the neighbouring water boundaries) and point sources (river inflow or direct discharges, if any). Other sources can be incorporated into the model if necessary, e.g., discharges from vessels.

Although this model was designed to calculate the fate of some PTS in 1995-2001, the simulation period can be extended if data requirements towards initial and boundary conditions are fulfilled. Furthermore, the model in this configuration can be applied to other shelf and regional seas.

III. Model Results and Discussion

Two compounds with different physical-chemical properties (Lammel et al., 2001) and different origins in the North Sea were selected for the modelling experiments: γ-HCH (an insecticide commonly known as lindane) and PCB 153 (a congener of polychlorinated biphenyls).

Lindane (γ-HCH) is a hexachlorocyclohexane isomer used as an insecticide in agriculture and forestry, as a wood and building preservative, and as a biocide to combat lice and scabies. The isomers α- and β-HCH occurred in the production of technical HCH, which was banned in 1980. The use of lindane is restricted, but despite this, α- and β-HCH are still widely found. A large proportion enters waterways through flooding, run-off from treated areas and incorrect disposal of leftover mixtures into farm drains and sewage systems.

PCBs have been used as coolants and lubricants in transformers, capacitors, and other electrical equipment, because they do not burn easily and are good insulators. The manufacture of PCBs was stopped in Europe in the end of the 1970s in light of evidence that they accumulate in the environment and can cause harmful health effects. Restrictions on the use of PCBs effected stoppage of their production in all countries surrounding the North Sea by 1985. For that reason, most inputs of PCBs occurred before 1980. PCBs are highly toxic, extremely persistent, and bio-accumulating. The greatest share of inputs of PCBs into the North Sea comes from the atmosphere. Sources of PCBs today are largely believed to be waste disposal and contaminated sites, especially non-regulated disposal of small capacitors that contain PCBs. PCB 153 is a congener of PCB and is often used as a PCB representative of all the others (an is often considered representative of the category as a whole), constituting around 10% of the total amount of PCBs.

Lindane (γ-HCH) is mostly observed in dissolved phase in water, whereas PCB 153 is less soluble and may well be adsorbed to suspended particles.

Although the North Sea is one of the better assessed seas in the world, measured data sets on PTS are not always easily accessible, have a very poor spatial coverage, and are limited to coastal monitoring stations. Moreover, the available values are sometimes unreliable, thereby introducing uncertainties to the model. Therefore, careful data selection is very important. The two selected compounds have sufficient data coverage to perform such a study. γ-HCH and PCBs are regularly monitored under the auspices of the OSPAR Convention and Bund-Länder monitoring programme (BLMP) for the marine environment of the North Sea.

Additional sources of uncertainties in the model are degradation rates in seawater, their dependence on ambient parameters (e.g., temperature and salinity), and a lack of process understanding. Thus, sensitivity analyses are an important means of evaluating PTS fate models.

FANTOM concentrations were compared with measurements available in different locations in the North Sea, presented here for one location in the German Bight for γ-HCH (figure 2). The model showed realistic spatial and temporal reproduction of the γ-HCH and PCB 153 concentrations for the years 1995-2001 in the North Sea. FANTOM is able to reproduce the general trend of decreasing γ-HCH concentrations in the North Sea for years 1995-2001 (figure 2). For PCB 153 there was no pronounced time trend predicted by FANTOM during the simulation period. This also agrees with observations. Datasets on PCB 153 occurrence in the North Sea are scarce making quantitative comparison of the model results with measured data not as promising as for γ-HCH.

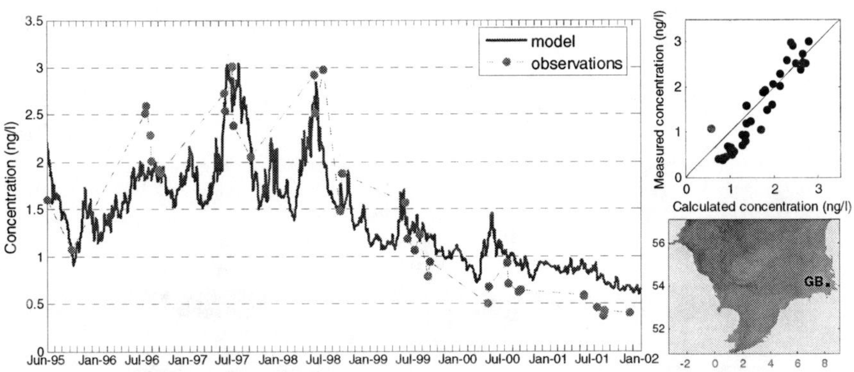

Fig. 2: Observed (dots) and FANTOM (solid line) γ-HCH concentrations (ng/l) in the surface layer in the North Sea plotted against time for the period 1995-2001 (left panel) and observed versus FANTOM γ-HCH concentrations (ng/l) at the same location (upper right panel). Location in the North Sea (in the German Bight, point GB) is shown on the map (lower right panel).

Annual mean γ-HCH and PCB 153 concentrations averaged vertically for the whole modelling domain in the years 1996–2001 were calculated and presented here for the beginning (1996) and the end (2001) of the simulation period (figure 3, see following page).

The North Sea is a flat marginal sea on the continental shelf in the North East Atlantic. It is closely linked to the Atlantic through manifold interactions. The Atlantic waters enter mainly in the north. Transport of pollutants, in this case lindane, follows the general circulation pattern. In the North Sea, distribution and mixing of water patterns are largely subject to meteorological conditions, run-off from rivers and the Baltic Sea, and tidal currents. Westerly winds prevail over the North Sea. Frequent observed circulation driven by winds and tides is anti-clockwise along the North Sea coast. However, during events of long lasting easterlies, the circulation pattern changes, bringing Baltic waters into the central North Sea, sometimes reaching all the way to the German Bight. Further inflow into the North Sea occurs via the English Channel. Run-off into the Atlantic occurs parallel to the western coast of Norway.

Both experimental and model results show steep gradients in γ-HCH and PCB 153 concentrations with higher values near the coast for dissolved and particle-bound fractions.

In 1996, the total concentration of γ-HCH in the continental coastal water was above 1 ng/l, with a concentration of more than 25 ng/l in the River Elbe estuary. γ-HCH concentrations in the whole modelling domain decreased. This effect is more pronounced in the southern regions of the North Sea where initial concentrations had been highest. Low concentrations in the north-western North Sea (figure 3) are due to the inflow of cleaner Atlantic water. Along the coastline from southern Britain to Denmark, close to the estuaries, there are high concentrations; this indicates the importance of the river inflow, which alters the local circulation.

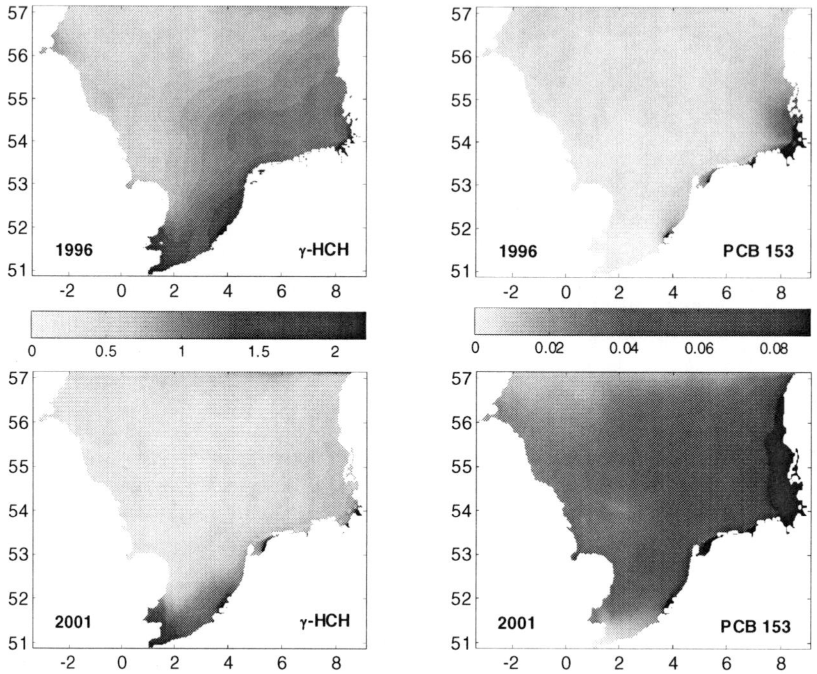

Fig. 3: Vertically integrated annual mean concentrations (ng/l) of γ-HCH and PCB 153 in the years 1996 and 2001 calculated by FANTOM.

The riverine inflow of γ-HCH, which is dominated by the inflow from the rivers Rhein and Elbe, exceeds the atmospheric deposition at the south-eastern coastal area of the North Sea. In contrast, in the central part of the North Sea the main source is atmospheric deposition.

PCB 153 concentrations decline offshore. This pattern continues throughout the entire simulation period. Concentrations of PCB 153 in 1995-2000 do not show any pronounced trend with the values between 0.025-0.05 ng/l in the southern North Sea and less than 0.01 ng/l in the central North Sea. In 2000 (and beyond in the model) PCB 153 levels increased in nearly the entire North Sea, with values higher than 0.035 ng/l measured (figure 3). This increase was also detected in tested biota, and in riverine and atmospheric samples.

Due to its high persistence and volatility, PCB 153 can be transported long distances in the atmosphere. An open question remains whether the PCB 153 seawater concentrations are still largely controlled by primary discharges or by re-emission from the sediments - a major repository of PCBs in aquatic environment. Recent studies favour the view that primary emissions control PCB 153 concentrations. Other contaminated water bodies such as the Great Lakes in North America have been reported to currently be net sources of PCBs to the atmosphere as a result of the release of PCBs accumulated in sediments. The mass balance studies performed for the Baltic Sea indicate that it is a net sink of PCBs. The phenomenon illustrated by figure 3 shows the significance of the sediment and atmospheric deposition contributions to the seawater PCB 153 burden.

IV. Conclusions and Outlook

Persistent Toxic Substances are harmful to human health and the environment. However, their fate and pathways in different environmental compartments are not yet fully understood. Models of PTS are recognised as helpful tools for assessing PTS cycling in natural systems. Model application is encouraged under the Convention on LRTAP and the Stockholm Convention. While new PTS are being produced and released into the environment, a modelling tool needs to be available to help in assessing their environmental fates. The role of oceans in general, and shelf and coastal seas in particular, as an exchanging compartment and/or permanent sink for POPs is yet to be fully understood. Furthermore, measurements of most PTS in seawater are still too limited (HCH isomers may be an exception) to allow for comprehensive assessment of their spatial and temporal distributions. Therefore, models have to be further developed and used to assess PTS sources and establish their environmental effects.

It is often thought that the lack of measurements on PTS is a major constraint for developing transport models. This study shows that sufficient data is available to perform a modeling study for the North Sea for some substances included in regular monitoring campaigns, such as HCH isomers and PCB congeners.

From this study, it has become clear that transport models can address the fate of some PTS, at least on the regional scale. Despite the uncertainties in some input parameters and lack of observational data, FANTOM was capable to reproduce realistic multi-year temporal and spatial distributions of γ-HCH and PCB 153 in the North Sea.

The modelling concept developed and applied in this study is also valid for simulating the fate of other contaminants in the North Sea that do not behave as passive tracers, e.g., radioactive substances or heavy metals. FANTOM can also be applied elsewhere such as the Baltic Sea or the Irish Sea.

In conclusion, it remains a future task to investigate the fate of contaminants with different environmental behaviour. This model could serve as the basis for further research on the contribution of different sources and sinks and sensitivity of contaminants' behaviour in the marine environment to the individual processes under present and future climate conditions.

References

Lammel, G., Feichter, J., Leip, A.: Long-range transport and global distribution of semi-volatile organic compounds: A case study on two modern agrochemicals. Report of the Max Planck Institute for Meteorology, No. 324, Hamburg, 44 pp, 2001.

Pohlmann, T.: Predicting the thermocline in a circulation model of the North Sea. Part I: Model description, calibration, and verification. Continental Shelf Research, 7, 131-146, 1996.

Sündermann, J., (ed.): Circulation and contaminant fluxes in the North Sea. Berlin, Heidelberg, New York: Springer, 654 pp, 1994.

Rebuilding the Eastern Baltic Cod Stock in a System of Change – An MPA Approach*

Christine Röckmann

I. Introduction

The Baltic cod stock is currently well below safe biological limits. Recent estimates of spawning biomass fall below B_{lim} (the precautionary biomass level, below which recruitment is impaired), implying the stock has a reduced reproductive capacity (ICES, 2003, 2004b). Consequently, the stock is classified "as being outside safe biological limits," posing concerns for Baltic Sea fisheries management (IBSFC, 2004).

The principal policy instrument for managing the Baltic Sea Fisheries is annual "Total Allowable Catches" (TACs), supplemented by technical regulatory measures, such as minimum landing sizes, mesh size regulations, and closed periods for fishing (IBSFC, 2002). Due to the persistently severe situation with respect to stock levels, the International Baltic Sea Fisheries Commission (IBSFC) in 2002 adopted a new recovery plan for the Eastern and Western Baltic cod stock. This recovery strategy focuses on meeting clearly defined stock reference points within several years (ICES, 2004b). In addition to the catch quotas, a spawning closure has been established since 1995, prohibiting cod fishing every year between 16 June and 15 August. The IBSFC also implemented a seasonal area closure on all fishing in the Bornholm Deep. This "summer ban" was extended by several weeks in 2005, lasting from 15 May until 31 August 2005. Moreover, the IBSFC established additional spawning area closures in the Gdansk Deep and the Gotland Deep. In spite of these new strategies, TACs remain in place, and are the main policy instrument. The effectiveness of TACs relies heavily on the quality of scientific fish stock assessments. These assessments are highly susceptible to the variability and uncertainty of a number of input parameters (e.g., catch-at-age and weight-at-age data, mortality estimates for the terminal years, environmental

* The results in this article are based on findings presented in detail in *Röckmann* et al., Testing the implications of a permanent or seasonal marine reserve on the population dynamics of Eastern Baltic cod under varying environmental conditions (2006), in press; <http://dx.doi.org/10.1016/j.fishres.2006.11.035.

parameters), and the difficulty in integrating complex processes, such as multi-species interactions and ecosystem functions (e.g., Botsford et al., 1997; ICES, 1998; Walters, 2001). Illegal landings also falsify the empirical data upon which the assessments are based.

Recently, there has been interest in the potential of marine reserves or marine protected areas (MPAs) as an alternative fishery management tool (e.g., Pauly et al., 2002; PROTECT, 2005; Roberts et al., 2001, 2005; Walters, 2001). Several field studies have illustrated that closed areas can lead to increases in fish biomass, density, and size, and ecosystem diversity (reviewed by Halpern, 2003). However, it is important to recognise the realistic limitations and expectations of the MPA approach, and to avoid presenting reserves as a general panacea (Allison et al., 1998; Dayton et al., 2000; Kaiser, 2005). Whether MPAs can enhance the fisheries outside of their borders is still in question, as success depends on the setting, size, and design of the reserve, the life history and migratory behaviour of the species, the specific characteristics of the ecosystem, and the fishing activity surrounding the MPA. In contrast to the tropics, there is less species variability in temperate waters where environmental conditions are much more variable, rendering the implementation of MPAs as fisheries management tools complicated. Few empirical studies have investigated the effects of MPAs on mobile temperate fish and their related fisheries. In temperate regions, modelling studies are currently a valuable tool to evaluate the implications of MPAs for migratory fish stocks and their respective fisheries.

This paper presents a spatially disaggregated, discrete time, age-structured model of the population dynamics of the Eastern Baltic cod (*Gadus morhua callarias L.*) to investigate the biological consequences of the establishment of a permanent or a seasonal marine reserve in a biologically highly productive area in the Baltic Sea. The model is a single species model, but is fed with output from an area-disaggregated Multispecies Virtual Population Analysis (adMSVPA). Instead of implementing the adMSVPA, here a simplified model is used, which compares favourably with the complex behaviour of the cod stock and the outputs of the adMSVPA. The advantage of this approach, relative to the adMSVPA, is its transparent structure, and its ability to couple it to complex bio-economic models, and thereby, to scenario-test the implications of management options and environmental conditions on stock dynamics.

The model was applied to simulate stock development of the Eastern Baltic cod over 50 years under various environmental conditions using different management policies. Management scenarios investigate the biological effects of the establishment of an MPA in SD 25 (Fig.1), which encompasses the Bornholm Basin, as opposed to policies based on the overall reduction of fishing mortality in the Eastern Baltic Sea. Marine biologists and conservationists have already recommended the establishment of an MPA in SD 25 (MacKenzie et al., 1996), as the Bornholm Basin has been the most important spawning ground of Baltic cod during the last decades due to favourable hydrographic conditions (Nissling and Westlin, 1991; Nissling et al., 1994; Plikshs et al., 1993).

II. Study Site Characteristics

Baltic cod eggs, which are neutrally buoyant, require a minimum salinity ($S \geq 11$ psu), and a minimum concentration of dissolved oxygen ($c[O_2] \geq 2$ ml/l) to develop (e.g., Nissling et al., 1994). The volume of water having these characteristics has been termed the "reproductive volume" (RV) for this species (e.g. MacKenzie et al., 2000; Pliksh s et al., 1993), and has been employed in the development of Baltic cod stock and recruitment models (e.g., Köster et al., 2001b, 2003; STORE, 2002). In the Eastern part of the Baltic Sea, a reproductive volume usually only occurs below a depth of 50 m in the deep basins of the Baltic Sea, namely, the Bornholm Basin, the Gdansk Deep, and the Gotland Deep, located in SD 25, 26 and 28, respectively (Fig.1) (Bagge et al., 1994). The bottom water in the deep basins regularly becomes anoxic due to bacterial degradation of organic material below the halocline, and reduced replenishment due to bottom topography resulting in restricted horizontal and vertical circulation in the Baltic deep water (Matthäus and Lass, 1995).

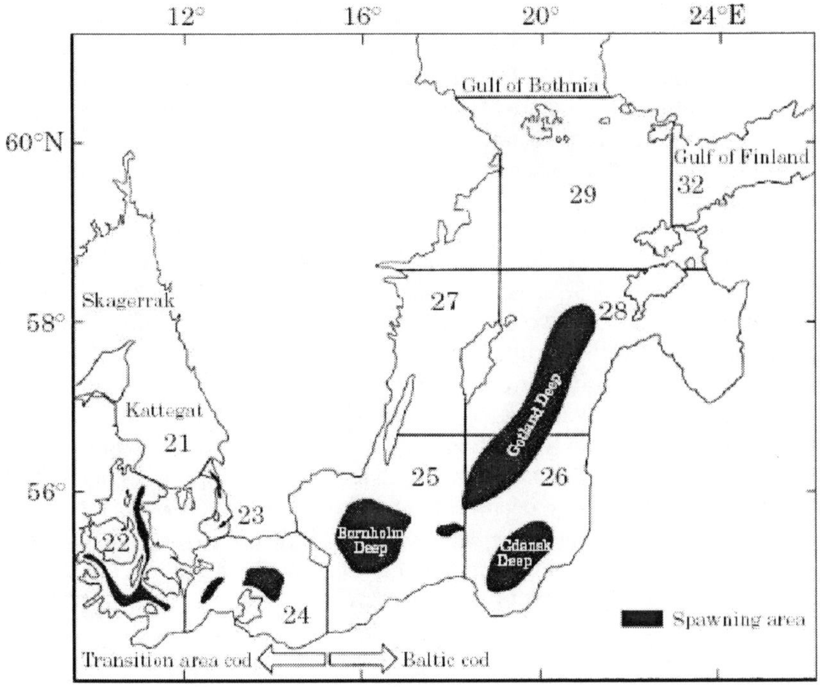

Fig. 1: Chart of the Baltic Sea, showing ICES subdivisions and important spawning grounds of Baltic cod. Source: after Bagge et al. (1994b).

The reproductive volume is replenished when inflows from the North Sea transport saline and oxygen rich water into the Baltic basins, termed "Major Baltic

Inflows" (MBI) (e.g., Matthäus and Frank, 1992; Schinke and Matthäus, 1998). Reproductive volume size depends mainly on the frequency and strength of North Sea inflow events as opposed to the length of stagnation periods (e.g., MacKenzie et al., 2000; Oeberst and Bleil, 2000). A large reproductive volume with hydrographic conditions favourable to successful cod egg development occurs less frequently in the Gotland Deep and the Gdansk Basin than in the Bornholm Basin. Their location farther East/Northeast, and hence more distant from the North Sea, leads to pronounced stagnation periods which prevail much longer than in the Western Baltic Sea (Köster et al., 2001a; Plikshs, 1993, 1996). Therefore, since 1986, the Gotland Deep in SD 28, and the Gdansk Basin in SD 26 have become less important for cod spawning. These facts support our hypothesis that a closure of ICES subdivision 25 will result in a significant biological benefit to the Baltic cod stock recovery.

III. The Model of Population Dynamics

The Eastern Baltic cod stock is distributed over ICES subdivisions (SD) 25-32 (Aro, 1989); however, the majority of cod are found in the three ICES subdivisions 25, 26, and 28 (Fig.1), to which this study is confined (Sparholt et al., 1991). The model is age-structured (age groups 0-8). Based on maturity estimates from maturity ogives, it is assumed that cod in age-group two and older are mature, and thus, are able to spawn (ICES, 2002; STORE, 2002). The age of entry into the exploitable fishery is also assumed to be two years.

Quarterly data from 1974 to 1999 on stock size, and natural, predation and fishing mortalities of cod for the ICES subdivisions (SD) 25, 26, and 28 from an adMSVPA was applied (ICES, 1999, 2001b; Köster et al., 2001a, 2001b). Basin-specific data on reproductive volume from 1976 to 1996 were specified based on estimates from MacKenzie et al. (2000) and Köster et al. (2001b), with estimates for the most recent years taken from ICES (2004a). An extension of the standard equation of population dynamics by Beverton and Holt (1954) was used to calculate cod stock size (N) for each age group (a), and for each subdivision (r), accounting for recruitment (R) and mortality due to fishing (F), predation (P), and natural mortality (M). For a detailed model description see Röckmann et al. (2005). The endogenous variables in the model are stock size (N), recruitment (R), and cannibalism (P). For parameterisation of the two sub-models calculating recruitment and cannibalism, multiple linear regression analyses was performed.

Cannibalism (Predation mortality P)

Predation mortality refers to cannibalism by mature cod on their early and juvenile life stages (ages 0, 1, and 2). In accordance with Köster et al. (2001b), predation mortality is linearly related to the cod spawning stock size (ssN), i.e., the sum of mature population numbers at ages 2-8 in the corresponding subdivision.

Recruitment (R)

In this study, "recruitment" refers to 0-group cod. If a spawning stocks exist, the young cod enter the model in the third quarter every year. From then on, the early life stages are subject to predation mortality (as reflected by cannibalism in our model), and to natural mortality. Recruitment in each of the three subdivisions is calculated as a function of the basin-specific spawning stock size (ssN) and the size of the basin-specific reproductive volume (RV). Several functional forms to combine the two explanatory variables ssN and RV were tested (Röckmann et al. 2005). A linear approach was chosen, which gave the best fit in SD 26 and 28 as well as good results in SD 25.

Fishing mortality (F)

The prime aim of this study is to analyze selected management policies, which constrain fishing mortality (F).Thus F is an external forcing parameter and treated as an exogenous variable. F differs between the investigated management scenarios, but is held constant during a management scenario. For the period of 1976-1999, the quarterly fishing mortalities derived by adMSVPA were applied. During the simulation period (2000-2050) the average fishing mortalities are modified according to the management policies (see below and Röckmann et al., 2005).

Natural mortality (M)

Corresponding to standard MSVPA runs in the Baltic Sea (Sparholt, 1991), natural mortality was assumed to be 0.2/year, equally distributed over quarters.

IV. Scenario Analysis

1. Management Scenarios

The analyses focus on the establishment of a marine reserve in SD 25. A comparison is made between the development of the stock size of Eastern Baltic cod under five selected management policies:

1. FasU — Fishing mortality "as usual," applying the average fishing mortality of 1990-1995 over the 50-year simulation period.
2. RoF70 — Overall reduction of fishing mortality by 70% in all three subdivisions (corresponds to ACFM advice for 2003).
3. C25q1q2 — Seasonal closure of SD 25 in quarters 1 and 2; quarters 3 and 4 are open to reduced fishing (fishing mortality in quarters 3&4 is reduced by 50%).
4. C25 — Permanent closure of SD 25.
5. TC — Total closure, i.e., fishing mortality is zero in all three subdivisions (corresponds to ACFM advice for 2005).

Model runs are performed covering the years 1976-2050 with the different management scenarios initiated in year 2000. Quarterly estimates of fishing mortalities are available from adMSVPA until 1999 (Köster et al., 2001a). Fishing mortality in the individual subdivisions after marine reserve implementation is derived from the pre-reserve fishing mortality, and the degree of fishing effort redistribution from SD 25 into SD 26 or 28. The effects of effort redistribution were not examined, as migration was neglected in this study.

2. Environmental Scenarios

Data on reproductive volume are available until 1999. To perform simulations of the Baltic cod population dynamics, future environmental conditions also need to be specified. The size of the reproductive volume in the three spawning basins depends on the frequency and strength of major Baltic inflows from the North Sea, which are triggered by large scale and local atmospheric forcing conditions, such as the North Atlantic Oscillation (Hinrichsen et al., 2002; Schinke and Matthäus, 1998). However, climate models projecting future atmospheric conditions are not yet available. This study does not attempt to predict reproductive volume; instead, the trends regarding cod stock development are examined based on the re-occurrence of historic environmental conditions. Thus, environmental scenarios for the simulation period (2000-2005) were based on the observed reproductive volume as it was estimated for the years 1974-1999. Under environmental conditions 1 ("HighLow"), the historic data set starting in 1974 was applied when reproductive volume was high in all three subdivisions. Under environmental conditions 2 ("LowHigh"), it was assumed that unfavourable conditions prevail in the first years of the simulation period; therefore, the simulation started with the reproductive volume data of year 1981, and then repeated the historic data sequence from 1981 through 1999, followed by 1974 to 1980. The two extreme cases of having very large ("HighHigh") or very small ("LowLow") reproductive volumes for a cycle of several years were also investigated.

V. Results

The modelled cod spawning stock size was compared with the spawning stock size derived from extended survivor analysis (XSA), i.e., the ICES standard stock assessment technique. Spawning stock size is underestimated by the model in the first part of the validation period (1976-1999), and overestimated towards the end of the validation period. These discrepancies can be attributed to flaws in the recruitment equation in SD 25, which does not accurately reproduce the observed recruitment, showing the same trends of underestimation in the late 1970s and early 1980s, and overestimation in the early 1990s. Nevertheless, the combined model used in this study explains 72% of the variance of ICES standard stock assessment estimates (2002) (Fig. 2).

The 50-year-simulations are initiated in the year 2000, when the size of the Eastern Baltic cod stock was very low, and the fishery was considered over-exploited. The curves in Figures 3-6 depict the trend of the development of the cod spawning stock in millions of fish, summed over SD 25, 26, and 28, for the four selected alternative environmental conditions.

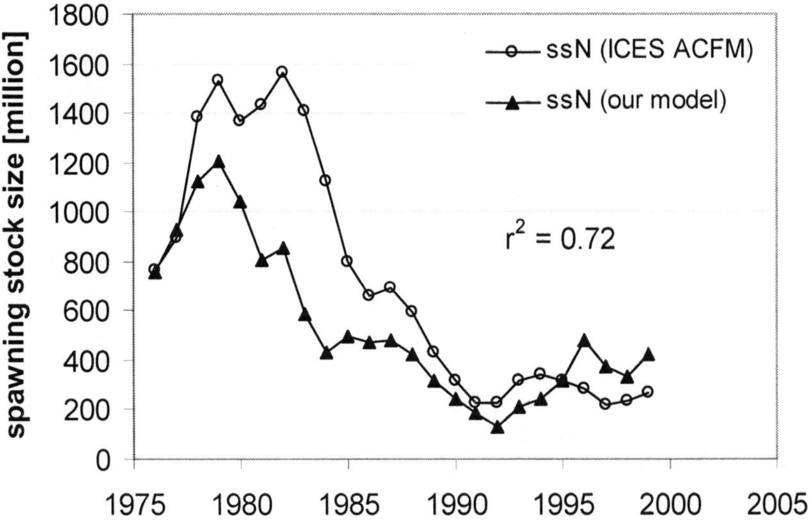

Fig. 2: Linear regression of the spawning stock size, derived by this model study with spawning stock estimates derived by ICES standard stock assessment; time series: 1976-1999.

The relationship between stock dynamics and reproductive volume under environmental conditions 1 ("HighLow") is described in detail in Figure 3. The simulation period under environmental condition 1 starts with several years of large reproductive volume in the three subdivisions (see the large black, white, and grey bars in Fig. 3). This leads to an erratic increase in the spawning stock size in 2004 under all five management scenarios (Fig. 3). The reproductive volume decreases within four years after the major Baltic inflow event in 2002, and remains low for about 10 years, from 2007 onwards. With a lag period of two years, the spawning stock size decreases gradually under all five management scenarios, while reproductive volume remains low. The decrease in stock size is most evident for management scenario 1 (FasU), whereas it is dampened in scenario 5 (TC) with fluctuations around a higher stock size. Under management scenarios 2, 3, and 4, the spawning stock size increases by 80, 100, and 120%, respectively, during the first seven years of the simulation period. Scenario 4 (C25) results in an initial increase in the cod spawning stock size to > 1 billion, a level similar to the spawning stock size at the end of the 1970s and early 1980s. Under management scenario 5 (TC), the spawning stock size increases by 160% to unprecedented high levels. Under management scenario 1 (FasU), the spawning stock size initially increases by 50% to approximately 700 million cod within four years. While reproductive volume is low, spawning stock size then decreases to less than 200 million spawners. This situation resembles the extremely low level of the spawning stock in 1992 (ICES, 2003, 2004b). The spawning stock does not go extinct under environmental condition 1, because the reoccurrence of inflows after about 15 years leads to the replenishment of reproductive volumes, and thus, to stock recovery.

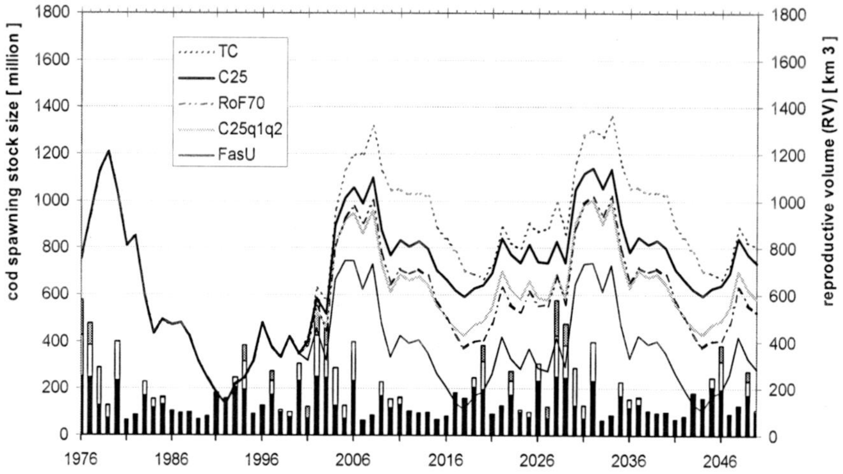

Fig. 3: Simulations based on environmental conditions 1 ("HighLow"): the data sequence from 1974-1999 is applied repeatedly for the simulation period 2000-2050. Management scenarios: ("NoC") no closure = fishing "as usual"; ("70C25") 70% closure of SD 25 & no fishing effort redistribution; ("100C25ER") 100% closure of SD 25 & full effort redistribution; ("100C25") 100% closure of SD 25 & no fishing effort redistribution; ("TC") total closure = zero fishing. Bars: reproductive volume [km^3] in SD 25 (black), in SD 26 (white), in SD 28 (grey).

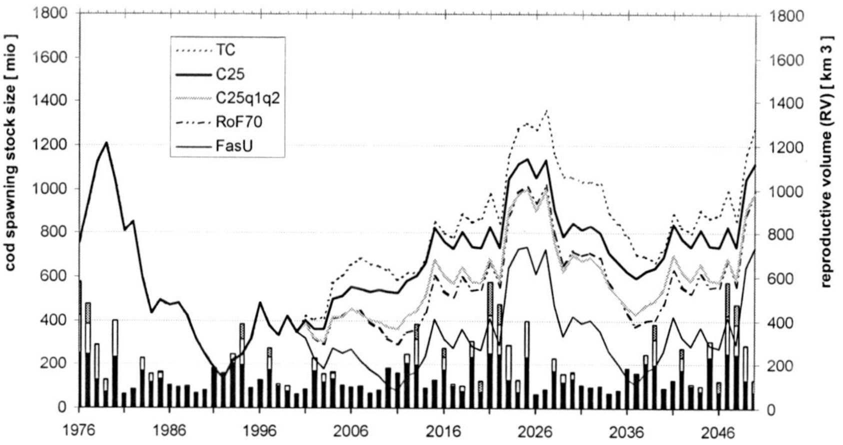

Fig. 4: Simulations based on environmental conditions 2 ("LowHigh"): the data sequence from 1981-1999 followed by 1974-1980 is applied repeatedly. See Figure 4 legend for explanation of the graphs.

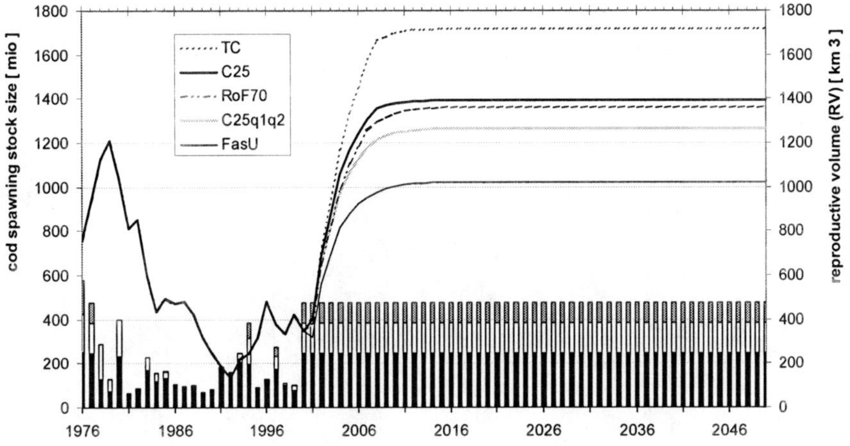

Fig. 5: Simulations based on environmental conditions 3 ("HighHigh"): constantly high reproductive volume; data from year 1977. See Figure 4 legend for explanation of the graphs.

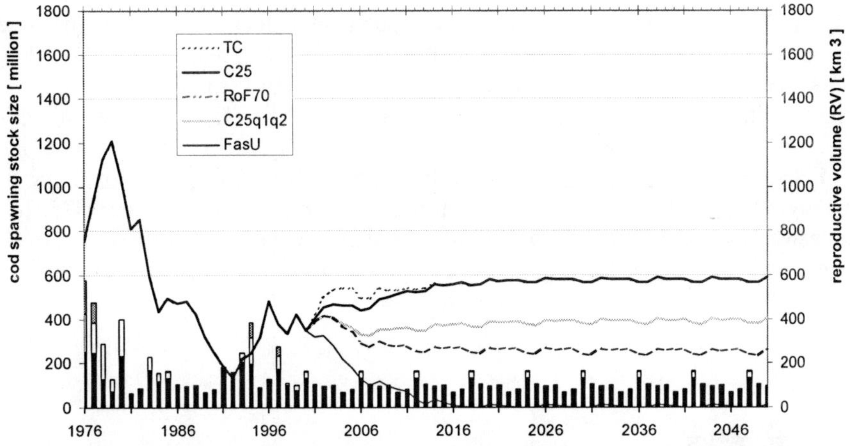

Fig. 6: Simulations based on environmental conditions 4 ("LowLow"): six year cycle of low reproductive volume; the data sequence from 1985-1990 is applied repeatedly. See Figure 4 legend for explanation of the graphs.

Figure 7 shows the age structure of the stock in 2050 under environmental condition 1 for the five selected management scenarios. Under the FasU scenario, hardly any cod grow older than 4 years. In contrast, under the total closure and the permanent MPA scenario, the stock's age structure improves greatly, supporting around 30% of 5- to 8-year old cod. Under the seasonal MPA scenario, this fraction of older fish contributes around 20% to the stock size. If overall fishing mortality is reduced by 70%, the stock is composed of approximately 10% of fish aged 5 and older.

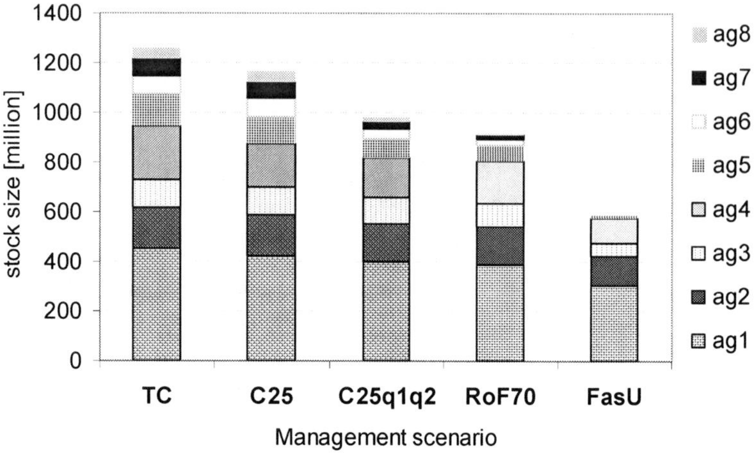

Fig. 7: Age structure of the cod stock in year 2050 for the five different management scenarios under environmental conditions 1 ("HighLow").

Figure 3 illustrates that in general, the spawning stock size under the five management policies differs to a greater extent during periods of stagnation (low RVs) than during and shortly after major Baltic inflows (high RVs). The lines representing the five management policies in Figure 3 lie closer together in years directly after major inflow events. Moreover, the C25q1q2 line is lower than the RoF70 line during years of high RV, but after a stagnation period of approximately 10 years, the two lines intersect, and spawning stock size under C25q1q2 exceeds that under RoF70. Considering the development of the spawning stock size towards the end of such a stagnation period, the ranking of management policies from highest to lowest spawning stock size is as follows:

1. TC, 2. C25, 3. C25q1q2, 4. RoF70, 5. FasU.

VI. Discussion

Management scenarios 2-5 represent different policy approaches for reducing fishing mortality. The simulation results clearly show that the reduction of fishing mortality via the establishment of a marine reserve in SD 25 is beneficial for the Eastern Baltic cod stock under both favourable and unfavourable environmental conditions (Figures 3-6). This is due to the fact that a reproductive volume will generally be present in the Bornholm Basin, despite unfavourable conditions, i.e., despite that lack of very strong inflow events from the North Sea, which could replenish the deep basins in SD 26 and 28. It is also evident that the cod stock could become virtually extinct if unfavourable environmental conditions prevail over a long time period in the future while current fishing levels are maintained (Fig. 6).

Management scenario 5 (TC) is probably the most unrealistic scenario from a management point of view; however, it corresponds to the recent management advice given to the IBSFC by the ICES advisory committee for fisheries management (ICES, 2004b). Due to the high uncertainties in the Baltic fish stock assessments, the ICES Advisory Committee on Fishery Management was reluctant to present a catch forecast in any form, and its management advice was: "no catch in 2005" (ICES, 2004b). Moreover, the total closure scenario (TC) can be interpreted as representing the environmental carrying capacity of cod in the Eastern Baltic Sea. According to our results, the carrying capacity ranges from approximately 600 million under unfavourable reproductive conditions ("LowLow") to 1.7 billion under the best case of reproductive conditions ("HighHigh"). Supposing that historic environmental conditions reoccur in the future, the carrying capacity varies between 1 and 1.4 billion, depending on the actual future reproductive conditions.

The simulations support the hypothesis that MPAs, both permanent and seasonal, can improve the age-structure of initially overexploited fish stocks (e.g., Apostolaki, 2001; Roberts et al., 2001), serving as positive feedback, particularly in the case of the Eastern Baltic cod (Fig.7). The advantage of having a greater number of older fish in the stock is that older females lay many more eggs and spawn over a longer time period than young females (Nissling et al., 1994, Vallin and Nissling, 2000). Moreover, the time period when the eggs are at maximum quality is longer with respect to the eggs of older females than for those of younger females.

The simulations also support claims raised by several fisheries scientists that the size and location of the marine reserve is a crucial factor for evaluating its impacts (Beattie et al., 2002; Martell et al., 2000; Sumaila, 2002; Walters, 2000). In the Baltic Sea, environmental conditions strongly drive the population dynamics of Baltic cod. Therefore, closing an area is useless if unfavourable hydrographic conditions do not allow for successful cod egg development, and in turn, do not increase potential recruitment. Closures lacking a sound biological basis can be counterproductive, as each negative example of marine reserve establishment can cause reluctance in fishermen to employ this type of management tool in the future. Furthermore, the Baltic cod is a migratory species with pronounced feeding and spawning migration. A small reserve size, such as the spawning ground closure established by the IBSFC in 1995, may dissipate any benefits because of dispersal losses or effort concentration around the borders of MPAs (Walters, 2000).

Similar to the results drawn by Brander and Mohn (2004), the simulations conducted in this study reveal that stock development depends strongly on future environmental conditions, particularly hydrological conditions triggered by interannual climate variability. There is a clear need to improve climate, meteorological, and hydrographic models which will allow for better fish stock projections, and provide scientifically sound advice to fishery managers and decision makers.

A major future goal in research is the integration of cod migration and movement rates. Tagging experiments are urgently needed to get quantitative information about cod migration.

VII. Conclusion

In summary, the results suggest that the Eastern Baltic cod stock can significantly benefit from the implementation of an MPA focusing on the protection of the spawning stock in SD 25. Under unfavourable environmental conditions, hydrographic conditions in the basins East/Northeast of SD 25 no longer allow for successful cod egg development. Therefore, it is particularly important the viable component of the spawning stock in SD 25 be protected. The stock will benefit more from a seasonal marine reserve policy than from an overall reduction of fishing mortality in the entire Eastern Baltic Sea.

In contrast to the investigated seasonal marine reserve scenario, the seasonal closure established in the Bornholm Basin is too small and too brief to effectively reduce spawning stock fishing mortality. Furthermore, the additional closures in the Gdansk Deep and the Gotland Basin may only be effective sporadically, in the rare case of strong MBIs. During stagnation periods, these closures cannot lead to enhanced recruitment from SD 26 and 28.

According to our findings, a total moratorium on fishing in the entire Baltic Sea could be avoided by an MPA approach focussing on SD 25 to rebuild the Eastern Baltic cod stock. MPAs represent a policy instrument that can, and should be applied in areas where the location of the stock's spawning concentration is known.

VIII. Acknowledgements

This study was funded by the International Max-Planck Research School for Maritime Affairs at the University of Hamburg and the Michael Otto Foundation for Environmental Protection. We also thank the EU project PROTECT for support.

References

Allison, G.W., Lubchenco, J., Carr, M.H., 1998. Marine reserves are necessary but not sufficient for marine conservation. Ecol. Applic. 8, pp. S79-S92.

Apostolaki, P., Milner-Gulland, E.J., McAlllister, M.K., Kirkwood, G.B., 2002. Modelling the effects of establishing a marine reserve for mobile fish species. Can. J. Fish. Aquat. Sci. 59, pp. 405-415.

Aro, E., 1989. A review of fish migration patterns in the Baltic. Rapports et Procès-Verbaux des Réunions du Conseil International pour l'Exploration de la Mer 190, pp. 72-96.

Bagge, O., Thurow, F., Steffensen, E., Bay, J., 1994. The Baltic cod. Dana 10, pp. 1-28.

Beattie, A., Sumaila, U.R., Christensen, V., Pauly, D., 2002. A model for the bioeconomic evaluation of marine protected area size and placement in the North Sea. Nat. Resour. Model. 15 (4), pp. 413-437.

Beverton, R.J.H., Holt, S.J., 1954. On the Dynamics of Exploited Fish Populations. Chapman & Hall, London.

Botsford, L.W., Castilla, J.C., Peterson, C.H., 1997. The Management of Fisheries and Marine Ecosystems. Science 277 (5325), pp. 509-515.

Brander, K., Mohn, R., 2004. Effect of the North Atlantic Oscillation on recruitment of Atlantic cod (Gadus morhua). Can. J. Fish. Aquat. Sci. 61 (9), pp. 1558-1564.

Chen, Y., 2003. Quality of fisheries data and uncertainty in stock assessment. Scientia Marina 67 (Suppl. 1).

Dayton, P.K., Sala, E., Tegner, M.J., Thrush, S., 2000. Marine reserves: parks, baselines, and fishery enhancement. Bull. Mar. Sci. 66, pp. 617-634.

Halpern, B.S., 2003. The impact of marine reserves: do reserves work and does reserve size matter? Ecol. Applic. 13 (1 Supplement February), pp. S117-S137.

Hinrichsen, H.-H., Möllmann, C., 2002. Biophysical modeling of larval Baltic cod (Gadus morhua) growth and survival. Can. J. Fish. Aquat. Sci. 59 (12), pp. 1858-1873.

Hinrichsen, H.-H., Lehmann, A., Mollmann, C., Schmidt, J.O., 2003. Dependency of larval fish survival on retention/dispersion in food limited environments: the Baltic Sea as a case study. Fish. Oceanogr. 12 (4-5), pp. 425-433.

Hinrichsen, H.-H., St. John, M., Lehmann, A., MacKenzie, B.R., Köster, F.W., 2002. Resolving the impact of short-term variations in physical processes impacting on the spawning environment of eastern Baltic cod: application of a 3-D hydrodynamic model. J. Marine Systems 32, pp. 281-294.

IBSFC, 2002. Fishery Rules of the International Baltic Sea Fishery Commission. http:// www. ibsfc.org/documentation/ibsfc_fishery_rules.

IBSFC, 2004. Proceedings of the thirtieth session. *Edited by* Ibsfc. Gdansk/Gdynia, Poland. 6.-10.9.2004. IBSFC.

ICES, 1998. Report of the Study Group on the Precautionary Approach to Fisheries Management. ICES CM 1998/ACFM:10 Ref.D.

ICES, 1999. Report of the Study Group on multispecies model implementation in the Baltic. ICES CM 1999/H:5, p. 199.

ICES, 2001. Report of the study group on multispecies predictions in the Baltic. ICES CM 2001/H:04.

ICES, 2002. Report of the Baltic Fisheries Assessment Working Group. ICES CM 2002/ACFM:17, pp. 1-547.

ICES, 2003. Report of the Baltic Fisheries Assessment Working Group. ICES CM 2003/ACFM:21, pp. 1-522.

ICES, 2004. Report of the Baltic Fisheries Assessment Working Group. ICES CM 2004/ACFM:22, pp. 1-522.

ICES, 2004. Answer to Special Request from International Baltic Sea Fishery Commission on areas within the Gotland Deep and Gdansk Deep where the hydrological conditions allow for a successful Cod spawning in 2004, published in IBSFC Proceedings of the thirtieth session (2004), p. 6.

Kaiser, M.J., 2005. Are marine protected areas a red herring or fisheries panacea. Can. J. Fish. Aquat. Sci. 62 (5), pp. 1194-1199.

Köster, F.W., Möllmann, C., 2000. Trophodynamic control by clupeid predators on recruitment success in Baltic cod? ICES J. Mar. Sci. 57, pp. 310-323.

Köster, F.W., Möllmann, C., Neuenfeldt, S., John, M.A.S., Plikshs, M., Voss, R., 2001a. Developing Baltic cod recruitment models. I. Resolving spatial and temporal dynamics of spawning stock and recruitment for cod, herring, and sprat. Can. J. Fish. Aquat. Sci. 58, pp. 1516-1533.

Köster, F.W., Hinrichsen, H.-H., John, M.A.S., Schnack, D., MacKenzie, B.R., Tomkiewicz, J., Plikshs, M., 2001b. Developing Baltic cod recruitment models. II. Incorporation of environmental variability and species interaction. Can. J. Fish. Aquat. Sci. 58, pp. 1534-1556.

Köster, F.W., Hinrichsen, H.-H., Schnack, D., St.John, M., MacKenzie, B.R., Tomkiewicz, J., Möllmann, C., Kraus, G., Plikshs, M., Makarchouk, A., Aro, E., 2003. Recruitment of Baltic cod and sprat stocks: identification of critical life stages and incorporation of environmental variability into stock-recruitment relationships. Scientia Marina 67 (Suppl. 1), pp. 129-154.

Kronbak, L.G., 2004. The Dynamics of an Open Access Fishery: Baltic Sea Cod. Mar. Res. Econ. 19 (4), pp. 459-480.

MacKenzie, B.R., St. John, M., Wieland, K., 1996. Eastern Baltic cod: perspectives from existing data on processes affecting growth and survival of eggs and larvae. Mar. Ecol. Prog. Ser. 134, pp. 265-281.

MacKenzie, B.R., Hinrichsen, H.-H., Plikshs, M., Wieland, K., Zezera, A.S., 2000. Quantifying environmental heterogeneity: habitat size necessary for successful development of cod *Gadus morhua* eggs in the Baltic Sea. Mar. Ecol. Prog. Ser. 193, pp. 143-156.

Martell, S.J.D., Walters, C.J., Wallace, S.S., 2000. The use of marine protected areas for conservation of lingcod (ophiodon elongatus). Bull. Mar. Sci. 66, pp. 729-743.

Matthäus, W., Frank, H., 1992. Characteristics of major Baltic inflows – a statistical analysis. Cont. Shelf Res. 12 (12), pp. 1375-1400.

Matthäus, W., Lass, H.U., 1995. The Recent Salt Inflow into the Baltic Sea. Journal of Phys. Oceanogr. 25, pp. 280-286.

Murawski, S.A., Brown, R., Lai, H.-L., Rago, P.J., Hendrickson, L., 2000. Large-scale closed areas as a fishery-management tool in temperate marine systems: the Georges Bank experience. Bull. Mar. Sci. 66, pp. 775-798.

Nissling, A., Westlin, L., 1991. Egg buoyancy of Baltic cod (*Gadus morhua*) and its implications for cod stock fluctuations in the Baltic. Marine Biol. 111, pp. 33-35.

Nissling, A., Kryvi, H., Vallin, L., 1994. Variation in egg buoyancy of Baltic cod *Gadus morhua* and its implications for egg survival in prevailing conditions in the Baltic Sea. Mar. Ecol. Prog. Ser. 110, pp. 67-74.

Oeberst, R., Bleil, M., 2000. Welche Faktoren beeinflussen die Stärke der Dorsch-Jahresklassen? Informationen für die Fischwirtschaft aus der Fischereiforschung 47 (1), pp. 31-37.

Pauly, D., Christensen, V., Guénette, S., Pitcher, T.J., Sumaila, U.R., Walters, C.J., Watson, R., Zeller, D., 2002. Towards sustainability in world fisheries. Nature 418, pp. 689-695.

Plikshs, M., 1996. Recent changes in cod spawning stock abundance and reproduction success in the Gotland area: is the cod recovery possible? ICES CM 1996/J:23.

Plikshs, M., Kalejs, M., Graumann, G., 1993. The influence of environmental conditions and spawning stock size on the year-class strength of the Eastern Baltic cod. ICES CM 1993/J:22 Sess.V, pp. 1-13.

PROTECT, 2005. Marine Protected Areas as a Tool for Ecosystem Conservation and Fisheries Management. EU 6th Framework Programme, Specific Targeted Research Project. Project No. SSP8-CT-2004-513670. http://www.mpa-eu.net/.

Reeves, S.A., 2001. The implications of age-reading errors for stock assessment and management advice: a case-study based on eastern Baltic cod. ICES CM 2001/P:18.

Ricker, W.E., 1958. Handbook of Computations for Biological Statistics of Fish Populations.

Roberts, C.M., Bohnsack, J.A., Gell, F., Hawkins, J.P., Goodridge, R., 2001. Effects of Marine Reserves on Adjacent Fisheries. Science 294, pp. 1920-1923.

Roberts, C.M., Hawkins, J.P., Gell, F.R., 2005. The role of marine reserves in achieving sustainable fisheries. Phil. Trans. R. Soc. B 360, pp. 123-132.

Röckmann, C., St. John, M.A., Köster, F.W., and Tol, R.S.J. 2005. Testing the implications of a marine reserve on the population dynamics of Eastern Baltic cod under varying environmental conditions. Working Paper FNU-63, Hamburg University and Centre for Marine and Atmospheric Science, Hamburg.

Schinke, H., Matthäus, W., 1998. On the causes of major Baltic inflows – an analysis of long time series. Cont. Shelf Res. 18, pp. 67-97.

Sparholt, H., 1991. Multispecies assessment of Baltic fish stocks. ICES Marine Science Symposia 193, pp. 64-79.

Sparholt, H., Aro, E., Modin, J., 1991. The spatial distribution of cod (*Gadus morhua* L.) in the Baltic Sea. Dana 9, pp. 45-56.

STORE, 2002. Environmental and fisheries influences on fish stock recruitment in the Baltic Sea. Final Consolidated Report (FAIR CT 98 3959), pp. 1-612 & 1-604.

Sumaila, U.R., 2002. Marine protected area performance in a model of the fishery. Nat. Resour. Model. 15 (4), pp. 439-451.

Vallin, L., Nissling, A., 2000. Maternal effects on egg size and egg buoyancy of Baltic cod, Gadus morhua: Implications for stock structure effects on recruitment. Fish. Res 49 (1), pp. 21-37.

Walters, C., 2000. Impacts of dispersal, ecological interactions, and fishing effort dynamics on efficacy of marine protected areas: how large should protected areas be? Bull. Mar. Sci. 66 (3), pp. 745-757.

Walters, C., 2001. Designing fisheries management systems that do not depend upon accurate stock assessment. *In* Reinventing Fisheries Management. Kluwer Academic Publishers, London, pp. 279-288.

Part II:

Prevention of Marine Pollution – Institutional Foundations

Prevention of Marine Pollution: The Contribution of IMO

Thomas A. Mensah

I. Introduction

The International Maritime Organization (IMO) is one of the specialized agencies of the United Nations. It was established by a constitutive Convention that was adopted in 1948 and entered into force in 1958.[1] The purposes of the Organization, as stated in Article 1(a) of the Convention, are "to provide machinery for cooperation among Governments in the field of governmental regulation and practices relating to technical matters of all kinds affecting shipping engaged in international trade; to encourage and facilitate the general adoption of the highest practicable standards in matters concerning maritime safety, efficiency of navigation and prevention and control of marine pollution from ships".[2]

In assessing the work of IMO for the prevention of marine pollution, it is pertinent to draw attention to two interesting facts concerning the relevance of marine pollution prevention to the mandate and work of the Organization. The first fact to be noted is that "prevention and control of marine pollution" was not among the purposes and functions of IMO when it was first established. Indeed, it was not until 1975 that specific reference to "the prevention and control of marine pollution from ships" was formally introduced into the constitution of the organization (the IMO Convention). The change was effected through an amendment to the IMO Convention adopted by the Assembly at its ninth session in 1975.[3] The

---

1 Convention on the Intergovernmental Maritime Consultative Organization (IMCO) 289 UNTS 3. The name of the Organization was changed to International Maritime Organization in 1982.

2 Convention on the International Maritime Organization (as amended in 1982), Article 1(a).

3 Amendments adopted by Resolution A.358(IX). The amendments entered into force in 1982. In addition to introducing "prevention of marine pollution from ships" as one of the purposes of the Organization, the 1975 amendments also "institutionalized" the Marine Environment Protection Committee (MEPC) and the Legal Committee as main organs of IMO. Until then the two committees were "subsidiary bodies" of the Assembly and Council, respectively. For a detailed dis-

second fact worth noting is that, although IMO is widely – and rightly - associated in the minds of academics, practitioners and environmentalists with the problem of marine pollution, the source of marine pollution that falls within its constitutional mandate (i.e. pollution from ships) is by no means the most important of the sources of pollution to the marine environment. Indeed, although incidents of marine pollution resulting from shipping casualties do attract considerable international attention, ship-borne substances represent less that a quarter of the sources of pollution of the marine environment.[4]

Notwithstanding these qualifications, the fact remains that IMO has an important role in the prevention of marine pollution and that it has, right from its inception, devoted a major part of its efforts and resources to the problem. In the process it has made important contributions to the international efforts to prevent, reduce and control pollution of the marine environment, especially pollution from vessels and by dumping. This contribution has been made largely through the development of a comprehensive technical, legal and administrative international regime for the prevention, control and reduction of pollution resulting from the routine operation of ships or from accidents involving ships carrying polluting substances. The regime developed by IMO comprises international regulations and standards as well as recommended practices and procedures applicable to vessels of all types. These regulations, standards and procedures are contained in international agreements, such as conventions and protocols, and also in non-binding instruments, including codes, recommendations and guidelines. In general the agreements and instruments set out legal, technical and administrative measures to be taken by States, individually and collectively, and also by private and public actors in the shipping industry. In establishing and operating its programme, IMO has also made an important contribution to the development and clarification of international environmental law, particularly international law regarding the protection and preservation of the marine environment.

When the idea of an international organization on shipping was first mooted, it was naturally expected that the new body would concentrate its efforts on what was considered to be its principle task, namely the improvement of safety and efficiency of maritime transport, with particular reference to the construction, equipment and operation of ships.[5] This was in fact what the new organization set out to do in the first decade of its existence. The first task undertaken by IMO was the adoption of a new version of the International Convention for the Safety of Life at Sea (SOLAS Convention), which still remains the most important international instrument on the safety of world shipping. The first version of the

cussion of the evolution of the mandate of IMO in the field of environmental protection see, generally, *M'Gonigle/Zacher, Pollution, Politics, and International Law* (1979) Chapter III.

4 It has for a long time now been accepted that the major source of pollution of the seas is pollution originating from land-based sources. See *UNEP*, The State of the Marine Environment (1990) GESAMP Reports and Studies No. 39. An Executive Summary of the Report is reproduced in *DOALOS*, Annual Review of Ocean Affairs (1993) Law and Policy Main Documents, at pages 250-3.

5 See *M'Gonigle/Zacher* (Note 3 supra).

SOLAS Convention adopted by IMO was the Convention of 1960. A revised and more comprehensive version was adopted in 1974.[6]

In the meantime, a new problem related to shipping was beginning to engage the attention of the international community. This was the problem of pollution from shipping operations. By the beginning of the 1950s, it had become apparent that action was needed to deal with the increasing pollution caused by the discharge of oil from ships operating at sea, and many countries found it necessary to introduce national regulations to control discharges of oil within their territorial waters. However, it soon became clear that the problem could not be dealt with effectively by individual and uncoordinated state action. International co-operation was needed.[7]

Actually, the process of harnessing international co-operation to deal with marine pollution from ships commenced sometime before IMO (or IMCO) commenced operations in 1959. In 1954 a conference was convened by the Government of the United Kingdom to consider measures to deal with the problem posed by discharges of oil and oily wastes from ships into the territorial seas of states. The 1954 conference adopted the International Convention for the Prevention of Pollution of the Sea by Oil, 1954, (the 1954 Oil Pollution Convention).[8] The depository and related functions under the 1954 Convention were assigned to the Government of the United Kingdom. but with the proviso that they would be transferred to IMO (IMCO), when it formally came into being and commenced operations.[9] The 1954 Oil Pollution Convention entered into force on 26 July 1958, soon after the entry into force of the IMCO Convention on 17 March 1958. At its first session in January 1959, the Assembly agreed that the depositary and related functions in respect of the 1954 Oil Pollution Convention should be exercised by the new organization; and these functions were duly transferred to IMCO by the Government of the United Kingdom. Thus it can be seen that IMO was associated with the prevention of marine pollution from ships even before it had come into being as a functioning institution.

The 1954 Oil Pollution Convention dealt primarily with pollution of the sea resulting from routine tanker operations, including the discharge of oily wastes from machinery spaces. At the time these were the major causes of pollution that were of concern to states. The 1954 Convention sought to tackle the problem in

---

6 International Convention on Safety of Life at Sea, 1974, 14 ILM 963 (1975). The Convention was further strengthened by a Protocol adopted in 1978.

7 On the developments leading to the establishment of IMO, see *M'Gonigle/Zacher* (Note 3 supra); also *de la Rue/Anderson,* Shipping and the Environment (1998) T.I.A.S 10009.

8 Adopted on 12 May 1954, 327 UNTS 3. This was the first (or one of the first) international treaty whose purpose was expressly and uniquely for the prevention of environmental pollution.

9 Article XXI of the Convention provided that "the duties of the Bureau shall be carried out by the Government of the United Kingdom of Great Britain and Northern Ireland until the Intergovernmental Maritime Consultative Organization comes into being and takes over the duties assigned to it under (the IMCO Convention)".

two main ways. First it established "prohibited zones", extending at least 50 miles from the nearest land within which discharges of oil or oily mixtures by ships were forbidden.[10] Secondly, the Convention urged all Contracting Parties to make every effort to provide facilities for the reception of oily water and residues in their ports. The idea was that the availability of reception facilities would decrease the temptation on crews to try to dispose of these residues at sea. The Convention was amended in 1962 and 1969.[11]

Although the 1954 Oil Pollution Convention went some way in preventing marine pollution from routine tanker operations, the rapid growth in the trade and transport in oil and related developments in industrial practices made it increasingly clear that further measures were required. However, there was not much enthusiasm for specific action among governments. At the time the international community did not appear to have been sufficiently alerted to the serious risks that the increase in the maritime transport of oil posed to the marine environment. This attitude was to undergo a radical change in the late 1960s, largely as a result of the *Torrey Canyon* accident in 1967. As a result of that incident the problem of marine pollution and its prevention moved much higher in the priorities of IMO and its member States.[12]

II. The Torrey Canyon incident

In the spring of 1967, the tanker *Torrey Canyon* ran aground while entering the English Channel and spilled her entire cargo of 120,000 tons of crude oil into the sea. This resulted in the biggest oil pollution incident ever recorded prior to that time. The incident raised serious questions not only with regard to the action that could be taken by a State when it was threatened with oil pollution as a result of a casualty involving an oil carrying tanker but also concerning the basis and procedures for compensating States or other entities which might suffer damage as a result of the casualty. In particular concern was expressed regarding the adequacy of the existing system of liability and compensation for such damage.

At the request of the Government of the United Kingdom an extraordinary session of the IMCO Council was convened to consider how IMCO - and the international community in general - should deal with the issues that had emerged as a result of the *Torrey Canyon* incident. The Council approved a "plan of action" which included the adoption of technical and legal measures to tackle these issues.

---

10 "Oily mixture" was defined as "a mixture containing more than 100 parts of oil per million".

11 The 1962 amendments extended the application of the Convention to ships of a lower tonnage and also extended the "prohibited zones" from 50 miles from the nearest land to 100 miles. The amendments of 1969 introduced regulations to further restrict operational discharge of oil from oil tankers and from machinery spaces of all ships. For more detailed information on the regime of the 1954 Oil Pollution Convention see IMO <www.imo.org/conventions/contents.asp/doc>.

12 On the implications of the *Torrey Canyon* accident see *Gold*, The Evolution of Marine Policy and Shipping Law (1981) Chapter 7.

A principal part of this plan of action was the convening of an international conference in 1973 to prepare a suitable international agreement that would place "restraints on the contamination of the sea, land and air by ships"[13].

Pending the 1973 conference, IMCO decided to introduce a series of measures intended to prevent tanker accidents or minimize the environmental consequences of any such accidents. Measures were also taken to deal with the increasing threat to the marine environment from routine tanker operations, such as the cleaning of oil cargo tanks and disposal of engine room wastes. It was noted that, in tonnage terms, these in fact presented a greater threat to the marine environmental than maritime accidents. For this purpose, the Assembly adopted further amendments to the 1954 Oil Pollution Convention in 1971.[14] The main purpose of the 1971 amendments was to limit the size of tanks on oil tankers, thereby minimizing the amount of oil that could escape in the event of a collision or stranding.

The general diplomatic conference envisaged by the Assembly was duly convened in 1973. The conference adopted the International Convention for the Prevention of Pollution from Ships, 1973 (MARPOL 1973).[15] The 1973 Convention was subsequently modified by a Protocol adopted in 1978.[16] The 1973 Convention, as modified by the 1978 Protocol, is referred to as MARPOL 1973/1978, and is widely recognized as the most important single treaty regime for the prevention of marine pollution from ship-borne substances. MARPOL 1973/1978 contains regulations for the prevention, reduction and control of accidental and operational pollution of the sea by oil as well by other substances carried on board ships, including chemicals, goods in packaged form and ship generated garbage and sewage. A later amendment has extended the scope of the Convention to the prevention of pollution of the air from ships.[17]

In addition to MARPOL, IMO developed two treaty instruments to address the legal questions that emerged following the Torrey Canyon incident. The first question concerned the measures which a coastal state could legitimately take to protect its interests when a maritime casualty threatened serious pollution damage to its territory, and the second related to the problem of liability and compensation when an incident caused pollution damage to a state or persons and entities within its territory.

With regard to the right of a coastal State to take measures to protect itself from environmental damage, IMO arranged for the preparation of a draft international convention which would prescribe the nature and limits of the measures that a coastal State could take in such circumstances. A diplomatic conference convened

---

13 IMO Council decisions at the third Extraordinary Session: IMO Doc.C/ES.III/5.

14 Text in 9 ILM 267.

15 Adopted 2 November 1973 Text in 12 ILM 1319;1340 UNTS 184 (1983).

16 17 ILM 546 (1978); 1340 UNTS 61(1983). Article 1 of the 1978 Protocol states that the provisions of MARPOL 1973 and the 1978 Protocol "shall be read and interpreted together as one single instrument".

17 Annex VI, adopted on 26 September 1977 and entered into force on 17 May 2005. The regulations in Annex VI set limits on sulphur oxide and nitrogen oxide emissions from ship exhausts and prohibited deliberate emissions of ozone depleting substances.

in 1969 adopted the International Convention Relating to Intervention on the High Seas in Cases of Oil Pollution Casualties, 1969.[18] This Convention affirmed that a coastal state had the right to take reasonable measures of intervention to prevent or minimize pollution damage to its coastline and related interests by oil escaping or discharged from a ship following an accident.[19]

On the issue of liability and compensation, two new conventions were developed by IMO. The first instrument, the International Convention on Civil Liability for Oil Pollution Damage (1969 CLC)[20], establishes a regime that makes it possible for States and persons who suffer pollution damage to obtain compensation more quickly than was previously possible. The Convention provides that the owner of the tanker from which oil has been discharged or escaped is directly liable to pay compensation for pollution damage resulting from the discharge or escape.[21] It also establishes a much simpler basis for this liability and introduces a new mechanism that ensures that compensation by or on behalf of the owner will be available in most cases. The second instrument adopted by IMO is the International Convention on the Establishment of an International Fund for Compensation for Oil Pollution Damage, 1971 (1971 Fund Convention).[22] This was described as "supplementary" to the 1969 Civil Liability Convention. Its purpose is to ensure that compensation will be available to victims of pollution damage when compensation under the 1969 Convention is insufficient to cover the damage caused or where, for some reason, the victim has been unable to obtain any compensation under the 1969 Convention.[23] The 1969 and 1971 treaties were amended in 1992, mainly to increase the level of compensation to be paid to victims of damage.[24]

III. Main areas of IMO's contribution to the prevention of marine pollution

The contribution of IMO to the international cooperation for the prevention of marine pollution has been in three broad areas. These are:

- the development of international regulations, standards and procedures;
- the establishment and operation of institutional mechanisms for the establishment, implementation and review of international rules and regulations; and
- the elaboration and clarification of international legal principles and norms.

---

18 Text in 9 ILM 24 (1970); 26 UST 765; T.I.A.S. 8068.
19 Article 1(1).
20 Adopted on 29 NOvember 1969, 9 ILM 45 (1970); 973 UNTS 3.
21 Article III(1).
22 Adopted 18 December 1971, 11 ILM 284 (1972).
23 Article 4(1).
24 Protocol of 1992 amending the 1969 Civil Liability Convention (1996 ATS No. 2; 1996 UKTS No. 86) and the Protocol of 1992 amending the 1971 Fund Convention (1996 ATS No. 3; 1996 UKTS No. 86).

IV. Adoption of international regulations, standards and procedures

The regulations, standards and procedures developed by IMO for the prevention, control and reduction of pollution of the marine environment may be categorized into three groups which are:

(a) regulations and procedures designed to prevent accidents involving ships which are carrying potentially polluting substances;
(b) regulations prescribing measures that should be undertaken after an accident has occurred, with a view to preventing or minimizing pollution from the substances on board the ship; and
(c) regulations specifying procedures and practices to be followed for the management and operation of ships, in order to prevent marine pollution from cargoes and other substances carried on board.

1. Regulations and procedures to prevent accidents to ships

A major part of the work of IMO for the prevention of marine pollution prevention has consisted in the adoption and revision of regulations, standards and procedures to prevent accidental discharge or escape from ships of potentially polluting substances. These regulations and standards are mainly in MARPOL 73/78 and include those relating to the design or construction of the vessels carrying oil and other substances. Among these are requirements intended to ensure that, in any loading condition, the vessel can survive damage by collision or stranding.[25] Some of the regulations require the fitting of segregated ballast tanks.[26] The availability of segregated ballast tanks enables the ship to keep ballast water away from tanks normally used for the carriage of oil as cargo. This reduces the need for the ship to have on board oily-water mixtures that the crew may feel compelled to dispose of at sea. The Regulations also mandate that segregated ballast tanks must be protectively located. In other words, they must be positioned in such a way that they will help protect the cargo tanks in the event of a collision or grounding of the ship. Another important design and construction provision of MARPOL is the requirement for tankers to have "double" bottoms or hulls.[27] This is to eliminate or reduce to the minimum the possibility that cargoes on board will escape or be discharged when the ship runs aground or is involved in a collision.

In addition to the requirements on the design and construction of the ship, there are regulations which oblige ships to carry on board special equipment considered necessary and useful for dealing with on board emergencies. The regulations have a two-fold purpose. First, they assist the crew to meet the requirements established

---

25 Regulation 22 of MARPOL 1973, as amended by the Protocol of 1978.

26 Regulation 14 of MARPOL 73, as amended by the Protocol of 1978. The 1978 MARPOL Protocol mandated segregated ballast tanks for all new tankers of 20,000 deadweight tons and above, as opposed to 70,000 deadweight tons in the 1973 Convention.

27 Amendments to Regulation 13F of MARPOL 73/78, as further strengthened by the amendments adopted in March 1992.

for the management of specified substances on board. Secondly, the presence on board of the equipment enables State and port authorities to ascertain if there has been compliance with relevant requirements of national and international law, when the ship calls at their ports. For example, tankers are required to have on board at all times oil-discharge monitoring and control equipment and oily water separating equipment and filtering systems [28] as well as oil discharge record books.[29]

The IMO regime also contains regulations which set out manning requirements for different types of ships. These are intended to ensure that the persons on board each ship are (a) fully aware of what is expected of them by national law and international treaties and (b) equipped professionally to discharge their responsibilities in all operating conditions. For example, the 1978 Convention on Standards of Training Certification and Watch-keeping of Seafarers (STCW)[30] sets out in its Chapter V special requirements to ensure that officers and ratings, who are entrusted with special duties relating to cargoes and cargo equipment on board, have all been given the requisite training. These requirements apply to oil tankers and also to chemical tankers and liquefied gas tankers.

The principal regulations, standards, and procedures for preventing accidental pollution from vessels, as contained in the relevant provisions of MARPOL 73/78 and in various amendments to the Convention adopted since 1978, are supplemented by numerous Codes, Recommendations and Guidelines adopted by IMO Assembly, the Marine Environment Protection Committee (MEPC) and the Maritime Safety Committee (MSC). In some cases, Codes that were originally adopted as non-binding recommendations have been made legally binding by being incorporated into the relevant treaties.[31]

The IMO regime for the prevention of accidental pollution is not confined to the provisions of the conventions and regulations that are expressly aimed at preventing marine pollution. Important and crucial components of the regime are to be found in various regulations and standards contained in conventions and other instruments whose primary purpose is to promote “safety of navigation”. By helping to reduce the possibility of accidents to ships carrying cargoes and substance that can cause pollution, the “safety” regulations and standards constitute an essential part of the programme to protect the marine environment from pollution from the various substances that are carried on board ships. Among the “safety” conventions are

(a) the 1974 SOLAS Convention and its amendments;
(b) the Convention on Load Lines, 1966;

---

28 Regulation 16 of MARPOL 73, as amended.

29 Regulation 20 of MARPOL 73, as amended.

30 Adopted on 7 July 1978, 1361 UNTS 190.

31 Examples are the amendments that made the International Code for the Construction and Equipment of Ships Carrying Dangerous Chemicals in Bulk (the IBC Code) and the Code for the Construction and Equipment of Ships Carrying Dangerous Chemicals in Bulk (BCH Code) mandatory under both MARPOL and SOLAS.

(c) the Convention for the Prevention of Collisions at Sea, 1972; and
(d) the Convention on Standards of Training, Certification and Watchkeeping of Seafarers, 1978.

As stated above, the conventions are supplemented by a large number of Codes, Recommendations and Guidelines adopted from time to time by the Assembly of IMO and the relevant technical committees.

2. Measures to deal with marine pollution emergencies

The States members of IMO recognize that it is not realistic to expect that the various measures adopted to prevent accidents to ships and to manage substances on board ships will all achieve a hundred percent success. They, therefore, accept that some accidents will happen, and that it is necessary and advisable to develop measures that States and relevant actors should take to prevent or reduce pollution of the marine environment when accidents do occur. For this purpose a number of international agreements and standards have been adopted by IMO. In general the agreements authorize or require States to take appropriate measures to deal with accidents which pose the threat of pollution to the marine environment. They also impose obligations on persons and various actors engaged in the operation and management of ships, including in particular shipowners and members of the crews of ships carrying potentially polluting substances. The obligations imposed by these convention relate, *inter alia*, to the construction, equipment and manning of ships, the handling of substances on board ships, the maintaining of records of operations at sea as well as the establishment of arrangements and procedures to anticipate accidents to ships or on board ships, and to deal with emergencies arising from such accidents.

With regard to measures to be taken by a coastal state to prevent or minimize pollution resulting from shipping incidents, IMO was called upon to deal with it as part of the response to the Torrey Canyon incident in 1967. At that time the main question considered was whether and to what extent a coastal state could take action to prevent or reduce damage to its interests in case of a maritime casualty. As noted earlier, that issue was directly addressed by the 1969 Convention Relating to Intervention in Cases of Oil Pollution Casualties.[32] The Convention expressly affirmed that a coastal state had the right "to take such measures on the high seas as may be necessary to prevent, mitigate or eliminate grave and imminent danger to its coastline or related interest from pollution or threat of pollution of the sea by oil, following upon a maritime casualty or acts related to such a casualty, which may reasonably be expected to result in major harmful consequences"[33] For a number of reasons, the application of the 1969 Convention was

[32] Note 18 supra.

[33] The right of the coastal state to take measures to protect itself from pollution is now expressly recognized as a principle of general international law of the sea. For example, it provides the basis of Article 9 of the 1989 Convention on Salvage which gives the right to a coast State to "give directions in relation to salvage operations" when a casualty threatens damage to its territorial sea.

restricted to incidents involving ships carrying oil, as defined in the Convention.[34] However, in 1973 a Protocol was adopted to extend the scope of the Convention to cover accidents involving vessels carrying potentially polluting substances other than oil.[35] A list of the substances within the scope of the Protocol is to be established and kept up to date by the Legal Committee of IMO, working in close cooperation with the Marine Environment Protection Committee (MEPC) of the organization.[36]

Subsequently, IMO found it necessary also to consider measures that States might be required to take to deal with maritime accidents that threatened to cause pollution to the marine environment in general. To ensure that States and other entities that may be affected would be adequately prepared to respond to emergencies that might arise from the casualties, IMO adopted the Convention on Oil Pollution Preparedness, Response and Cooperation, 1990.[37] The purpose of this Convention is to ensure that coastal states are able, prepared and suitably equipped to take the measures necessary to prevent, mitigate or avoid serious pollution from maritime casualties. Parties to the convention are required to establish measures and arrangements for dealing with pollution incidents, either individually or in cooperation with other countries. In particular, they are required to establish stockpiles of oil spill combating equipment, to hold oil spill combating exercises and to develop detailed plans for dealing with oil pollution emergencies.[38]

The 1990 Convention also imposes special obligations on operators of ships. Among others, the Convention makes it mandatory for tankers to carry shipboard oil pollution emergency plans, and these plans are required to be co-ordinated with national systems for responding promptly and effectively to oil pollution incidents.[39] The master and crew of ships are also required to report to the appropriate coastal state any incident of oil pollution involving their ship.[40]

3. Measures and procedures to prevent pollution resulting from operations on board ships

In addition to measures and prescriptions to prevent accidental pollution from ships, IMO has developed a number of standards and practices to eliminate and

---

34 Oil was defined in the Convention as "crude oil, fuel oil, diesel oil and lubricating oil" (Article 1, paragraph 3. The conference decided not to include whale oil in the scope of application of the Convention.

35 Protocol Relating to Intervention on the High Seas in Cases of Marine Pollution by Substances Other than Oil, 1973 UKTS 27 (1983).

36 Article III of the 1973 Protocol.

37 Adopted 30 November 1990 Text in 30 ILM 773 (1991). A Protocol was adopted in 2000 to extend the scope of the Convention to pollution emergencies involving substances other than oil (Protocol on Preparedness, Response and Co-operation to pollution Incidents by Hazardous and Noxious Substances, 2000 (OPRC-HNS Protocol).

38 OPPC Convention, Article 6.

39 OPPC Convention, Article 3.

40 OPPC Convention, Article 4.

reduce the possibility of damage to the marine environment from activities undertaken on board ships. These include practical procedures and practices to be adopted by oil tankers and other ships that have on board substances that can cause pollution of the sea. Examples are:

(a) the system of "load on top" for the carriage of ballast water. Under this system oil taken on board a tanker as new cargo is loaded "on top" of oily water mixtures that may be in the cargo tank from a previous voyage. The oily water mixture can thus be retained in the tanks until it can safely be discharged into appropriate reception facilities at the next port of call. This avoids the need and temptation for the crew to dispose of the oily-water mixture at sea. [41]
(b) Another procedure introduced to prevent "operational pollution" by oil was the method of "crude oil washing" by which oil tanks are cleaned not with sea water, as previously, but rather with crude oil. The dirty oil resulting from the cleaning is kept in the tank, along with the new cargo taken on board. This means that there is no mixture of oil and water which the crew might be tempted to get rid of at sea.[42]
(c) Another measure adopted to avoid pollution from shipping operations is the requirement for states to provide for reception facilities in ports. The availability of reception facilities at ports provides an incentive for the crew to retain any oily-water mixtures that may be on board until they reach the next port since they can confidently expect that they will be able to get rid of these at the port and make room for new cargo to be taken on board. [43]

A major part of the MARPOL regime is the section dealing with the designation of "special areas" where discharges of oil or oily-water mixtures is completely prohibited.[44] The designation of special areas, coupled with the requirement that tankers should keep records on board for inspection at ports, provides a very

41 For a detailed discussion on the Load on Top (LOT) system, see *M'Gonigle/Zacher* (Note 3 supra) 96-114, passim.

42 "Crude oil washing is a process whereby the washing of cargo tanks prior to loading ballast water is accomplished by using the crude oil cargo itself as a washing agent during cargo unloading operations. When crude oil is uses as a washing agent, it is many times more efficient than water because the oil acts as a solvent, dissolving the sludge and sedimentation, which is then pumped ashore" *de la Rue/Anderson* (Note 7 supra) 766, note 52. In the 1978 MARPOL Protocol, Crude Oil Washing (COW) was accepted as an alternative to the requirement for segregated ballast tanks (dedicated ballast tanks) for certain ships.

43 While the 1954 Oil Pollution Convention merely urged Parties to make provision for reception facilities in their ports, MARPOL 73/78 makes the provision of such facilities mandatory on all Parties to the Convention, Regulation 12.

44 MARPOL 73 designated the Mediterranean Sea, the Black Sea, the Baltic Sea, the Red Sea and the Gulfs Area as "special areas". Since then several other areas have been designated as special areas, including the Oman Sea Area, and the North West European Waters Area.

powerful incentive for members of the crew to desist from discharging oily-water mixtures into vulnerable areas of the sea.

V. Establishment and operation of institutional mechanisms

The bulk of IMO's work in the field of marine pollution prevention is undertaken in the specialized institutional mechanism provided by the governing organs and technical committees. These organs and specialist committees provide the forum in which States, international organizations, private bodies and institutions in the maritime industry and various stakeholders in the environmental movement consider and adopt regulations, standards and recommended practices and procedures of all kinds for the prevention of marine pollution prevention. They also constitute the institutional mechanism by which the regulations and standards are reviewed and up-dated, to ensure that they respond effectively and appropriately to new development and needs. The most important organs are the Council and Assembly; and the most significant specialist committees are the Marine Environment Protection Committee (MEPC), the Maritime Safety Committee (MSC) the Consultative Meetings of Parties to the London Dumping Convention; and the Legal Committee.

1. The Marine Environment Protection Committee (MEPC)

The Marine Environment Protection Committee was established in 1973 as the principal inter-governmental organ to co-ordinate the work of IMO in the prevention and control of marine pollution prevention from ships.[45] In 1985 the Committee was upgraded and constituted as one of the main organs of the IMO.[46] Through the MEPC and other subsidiary technical committees, IMO has adopted conventions on the prevention of marine pollution from accidents to ships and from routine operations of ships. The most important of these conventions is the 1973/78 MARPOL and the various amendments that have been adopted since 1978. MARPOL is supplemented by a large body of Codes and Recommendations intended to clarify and implement the provisions of the conventions.[47] The MEPC has also established Guidelines on how various provisions of MARPOL and its related conventions are to be implemented in practice.[48] In some cases the MEPC has issued "agreed interpretations" of provisions of the MARPOL whose meaning might otherwise be the subject of controversy between different states or groups of

---

45 Resolution A. 297 (VIII).

46 By resolution A. 358(IX) Note 3 supra.

47 Such as, the International Dangerous Goods (IMDG) Code; the Code for the Construction and Equipment of Ships Carrying Dangerous Chemicals in Bulk (IBC) Code and the International Safety Management (ISM) Code.

48 For example, Guidelines for the Transport of Vegetable Oils in deep Tanks or in Independent Tanks specially designed for the Carriage of such Vegetable Oils on board of Dry cargo ships (MEPC, 52nd Session, October 2004).

states. In addition to its work relating to the implementation and review of MARPOL, the MEPC has adopted other conventions or guiding resolutions on specific aspects of marine pollution prevention, including those relating to the treatment of ballast water,[49] anti-fouling systems[50] and ship recycling etc.[51]

2. Maritime Safety Committee (MSC)

The Maritime Safety Committee has, from the inception of IMO, been the premier technical body of the Organization. Its man function is to develop rules, regulations, standards and procedures relating to maritime safety and efficiency of navigation. In that capacity the MSC has been instrumental in establishing a maritime safety regime contained in conventions such as the 1974 SOLAS, the 1966 Convention on Load Lines, the 1978 Convention on Standards of Training Certification and Watchkeeping of Seafarers (STCW), the 1972 Convention on the Prevention of Collisions at Sea as well as numerous Codes and Recommendations relating to the construction, equipment and operation of large variety of ship types and other craft.

3. The Consultative Meeting of Contracting Parties to the London Dumping Convention (LDC)

The Consultative Meetings of Parties to the London Dumping Convention[52] is operated and administered by IMO, at the request of the States Parties to the 1972 London Dumping Convention. IMO was designated by the States Parties to perform the secretariat functions in relation to the convention.[53] In response IMO has provided secretariat and back-up support for the Consultative Meetings in the discussions relating to the implementation and up-dating of the regime of the 1972 London Dumping Convention. The work of the Consultative Meetings has resulted in several important revisions designed to improve the Convention. Among these are the amendments to the Convention that ban incineration of industrial wastes at sea and prohibit the dumping at sea of low-level radioactive wastes.[54]

---

[49] International Convention for the Control and Management of Ships Ballast Water Sediments, adopted 13 February 2004.

[50] International Convention for the Control of Anti-fouling Systems in Ships, adopted on 30 October 2001.

[51] For example, the IMO Guidelines on Ship Re-cycling (Assembly resolution A. 962(23).

[52] 11 ILM 1294 (1972). The Consultative Meetings of Parties to the Convention are convened pursuant to paragraph 3(a) of Article XIV of the Convention.

[53] Pursuant to Article XIV(2), which requires the Contracting Parties (at their first meeting) to "designate a competent Organization ... to be responsible for Secretariat duties in relation to (this) Convention". The Secretariat duties referred to are set out in detail in Article XIV(3).

[54] These amendments were adopted in November 1993. These methods of disposal were permitted under the original Convention. The amendments were preceded by

Perhaps the most important result of the work of the Consultative Meetings of Parties has been the adoption in 1996 of a Protocol that is intended to replace the original convention.[55] In addition to entrenching the "precautionary approach" and the "polluter pays principle" firmly in the regime of the convention, the 1996 Protocol adopts a more restrictive approach in the regulation of dumping. Thus, instead of the old approach under which Contracting States had the right to authorize the dumping of various categories of substances and wastes, the 1996 Protocol requires States Parties to prohibit the dumping of "any wastes or other matter with the exception of those listed in Annex 1".[56]

4. The Legal Committee

The Legal Committee is one of the main committees of IMO. It was first established in 1967 as an *ad hoc* body to deal specifically with the legal issues that arose in the aftermath of the Torrey Canyon incident. It was later made a subsidiary organ of the Council. Subsequently, it was institutionalized as one of the "organs" of IMO by 1975 amendments to the IMO Convention which entered into force in 1982.[57] The Legal Committee has been responsible for the development of several treaty instruments dealing with legal issues, including in particular the 1969 Convention Relating to Intervention on the High Seas in Cases of Oil Pollution Casualties (and its 1973 Protocol);[58] the 1969 Convention on Civil Liability for Oil Pollution Damage and its Protocol of 1992;[59] the 1971 Convention on the Establishment of an International Fund for Compensation for Oil Pollution Damage and its Protocol of 1992;[60] the 1989 Convention on Salvage[61] and the 1996 Convention on Liability and Compensation for Damage in Connection with the Carriage of Hazardous and Noxious Substances by Sea (HNS)[62] The Legal Committee performs important functions relating to the implementation of some of these Conventions. For example, the Committee has the mandate to prepare and adopt amendments to revise the limits of compensation payable to victims of pollution damage covered by the 1992 Civil Liability, the 1992 Fund Convention[63] and the 1996 HNS Convention.

---

resolutions of the Consultative Meetings calling for a moratorium on these practices.

55 26 ILM 7 (1997).

56 1996 LDC Protocol, Article 4(1.1).

57 Resolution A. 358(IX), Note 3 supra.

58 Note 18 supra.

59 Note 20 supra.

60 Note 22 supra.

61 Text in IMO document LEG/CONF.7/27 of 2 May 1989.

62 35 ILM 1406 (1996).

63 See Article 15 of the 1992 Civil Liability Convention; and Article 33 of the 1992 Fund Convention.

VI. Progressive Development of International Law

In the course of its work for the prevention of marine pollution from ships, IMO has also made important contributions to the implementation of the United Nations Convention on the Law of the Sea and to the development of general environmental law, at both the national and international levels. In the some cases the results of the work of IMO have been considered to be necessary for the practical implementation of some provisions of the United Nations Convention on the Law of the Sea. As is well known, the 1982 Convention calls for the establishment of international rules and standards to for the prevention, reduction and control of pollution of the marine environment from vessels; and also enjoins and authorizes States to adopt laws and regulations for the same purpose. However, the Convention requires that the laws and regulations adopted by States should be compatible with the Convention and other rules of international law. The Convention further states that these laws and regulations should "at least have the same effect" as that of "generally accepted international rules and standards established through the competent international organization".[64] This means that, in some cases, the benchmark by reference to which laws and regulations adopted by a State are evaluated is whether those laws and regulations are compatible with, or have the same effect as, the relevant regulations and standards established from time to time by IMO. Thus, the regulations and standards developed by IMO are useful not only for the practical implementation of some provisions of the Convention but also for evaluating the laws and standards enacted by States to implement particular provisions and rules of the Convention. The areas in which the results of the work of IMO are necessary or useful either for the implementation of the Convention or the evaluation of national laws and regulations include the following:

(a) Evaluation of coastal states of laws and regulations adopted pursuant to article 21 of the Convention, especially those with respect to safety of navigation and the regulation of maritime traffic; the protection of navigational aids and facilities and other facilities or installations; and the preservation of the marine environment and prevention of pollution, reduction and control of pollution thereof. According to Article 21 of the Convention, such national laws must be in conformity with the provisions of the (present) Convention and other rules of international law. In particular, these laws and regulations shall not apply to the design, construction, manning or equipment of foreign ships "unless they are giving effect to generally accepted international rules and standards". In respect of the design, construction, manning or equipment of ships, the only "generally accepted" rules and standards are those that have been developed and adopted in IMO. These are contained in the SOLAS Convention, the International Convention on the International Regulations for Preventing Collisions at Sea, 1972, the 1973/1978 MARPOL

---

[64] Convention on the Law of the Sea, Article 211, paragraphs 1 and 2. It is generally agreed the "the competent international organization", as used in the Convention, refers to IMO.

Convention, the STCW Convention and in the numerous Codes, Recommendations and Guidelines related to these treaty instruments.[65]

(b) The establishment by coastal states of traffic separations schemes in territorial seas, archipelagic waters or in straits used for international navigation. Articles 22, 41 and 47 of the Convention provide that coastal or archipelagic states which designate sea lanes or prescribe traffic separation schemes shall take into account, *inter alia*, the recommendations of IMO ("the competent international organization").

(c) Article 23 of the Convention provides that foreign nuclear powered ships and ships carrying nuclear or other inherently dangerous substances are required, when exercising the right of innocent passage through the territorial sea, to carry documents and observe special precautionary measures established for such ships in international agreements. Many of the internationally accepted regulations and standards for the maritime carriage of noxious and dangerous substances are contained in the International Dangerous Goods Code (IMDG Code) which has been developed largely in IMO in cooperation with other United Nations bodies including the International Atomic Energy Agency (IAEA).[66]

(d) Pursuant to Article 210 of the Law of the Sea Convention (prevention of pollution by dumping) and Article 221 (prevention of pollution from vessels), international rules and regulations for the prevention of pollution shall be established through "the competent international organization" (i.e. IMO). The articles also provide that national laws adopted for the same purpose should at least "have the same effect" as the international rules and standards established through IMO ("the competent international organization").

In addition to providing standards and parameters for evaluating international and national laws and regulations for the prevention, control and reduction of marine pollution, IMO has also made an important contribution to the progressive development of international law for the protection and preservation of marine environment, in general, and the prevention of vessel source pollution, in particular. For example, IMO has contributed significantly to the articulation and practical implementation of a number of "accepted" or "general" principles and norms of law relating to the protection of the environment. Indeed, in some cases, the principles were first enunciated in IMO while, in other cases, the principles have been given practical application in instruments developed in IMO. Among these the following may be mentioned:

---

65 See *Shabtai Rosenne*, IMO Interface with the Law of the Sea Convention, in: Current Maritime Issues and the IMO, ed. by *Nordquist/Moore* (1999) 251-268.

66 IMO co-operates with a number of United Nations agencies and bodies of the United Nations in the work of the Group of Experts on Scientific Aspects of Marine Pollution (GESAMP). A great deal of the work relating to the IMDG Code takes place within the framework of GESAMP for which IMO provides the Secretariat.

(a) The principle that a coastal state is entitled to take measures of intervention to protect itself in cases of serious maritime accidents was first articulated in express terms in the 1969 Convention Relating to Intervention on the High Seas in Cases of Oil Pollution Casualties that was adopted in IMO following the Torrey Canyon accident.[67] As indicated earlier, this principle is now enshrined as a general rule of international law of the sea by article 221 of the 1982 UN Convention on the Law of the Sea.

(b) The rule that a port State may take measures in respect of a ship in its port or off-shore terminal, either to ascertain compliance with applicable international and national laws rules and regulations or to take enforcement action in case of proven violations, was proclaimed in the MARPOL Convention of 1973, well before the concept of "port state jurisdiction" to enforce national and international law against foreign vessels was formally promulgated as a general principle in articles 218 and 220 of the Convention on the Law of the Sea.[68] This principle has provided the principal legal basis for various Memoranda on Port State Control now in operation in many areas of the world.[69]

(c) The principle that a port State can apply the requirements of a treaty to which it is a Contracting Party to a foreign ship in its port, even when the flag State of the ship has not accepted that particular treaty, was recognized in IMO conventions before that rule was adopted in the United Nations Convention on the Law of the Sea. That principle was first applied in the 1969 Civil Liability Convention. Article VIII of that Convention provides that each Contracting State shall ensure that insurance or other security to the extent required under the Convention is in force in respect of any ship, wherever registered, entering or leaving its ports or off-shore terminals in its territory. The same principle is applied in Article 5, paragraph 4, of the 1973 MARPOL Convention. That provision imposes an obligation to apply the provisions of the Convention on the same terms to all ships.[70] The principle behind these provisions is now accepted as an integral part of effective international environmental law. However, the principle was strongly resisted by many governments when the provision was first proposed in IMO. These governments argued that such a provision would be incompatible with the general international law rule that a treaty cannot impose obligations, on or

---

67 Note 18 supra.

68 Article 5, paragraph 2, of M\ARPOL73/78 provided that" a ship required to hold a certificate in accordance with the provisions of Regulations (of MARPOL) is subject, while in the port or off-shore terminal under the jurisdiction of a Party, to inspection by officers duly authorized by that Party ...".

69 See, for example, the Paris Memorandum of Understanding and the Tokyo Memorandum of Understanding (Memorandum of Understanding on Port State Control in the Asia-Pacific Region, Tokyo, 1 December 1993).

70 Article 5(4) of MARPOL provides that the requirements of the Convention shall be applied to ships of non-parties "as may be necessary to ensure that no more favourable treatment is given to such ships".

adversely affect the rights of, a third party State i.e. a State which has not consented to be bound by the treaty.

(d) The use of the "tacit acceptance" procedure to adopt urgently needed amendments to international environmental agreements was pioneered by IMO. Prior to this, amending such agreements involved considerable difficulty and delay because of the prevailing view that a contracting State could only be bound by an amendment if it had explicitly expressed its consent to be bound by that amendment. The "tacit acceptance" procedure was promoted and used in IMO because many States were unhappy that necessary amendments to international conventions could not be brought into force in a timely manner. In many cases, much needed amendments could not be implemented in time or at all by means of the "traditional" procedures of the international law of treaties. These procedures required first, that the amendment should be ratified or accepted by a stipulated proportion (usually two-thirds) of the States Parties and, secondly, that upon entry into force the amendment would be binding only on the States Parties which had accepted it. The latter requirement meant that, even when an amendment came into force, it would not be applicable to any of the States Parties to the treaty that had not expressly agreed to be bound by the amendment. A consequence of these requirements was that, as the number of parties to a treaty increased, so did the number of acceptances required to bring an amendment into force. In effect it was like "trying to climb a mountain that was always growing higher"[71] The problem was made more difficult by the fact that Governments took much longer to accept an amendment than they did in ratifying the parent convention. Following extensive discussions in the legal and technical committees of IMO, it was decided to adopt a new and more expeditious procedure for amending the technical regulations and standards in IMO's conventions. The new approach was approved by a resolution of Assembly in 197.1[72] The Assembly noted that there was need for an amendment procedure for IMO treaty instruments that would be "more in keeping with the development of technological advances and social needs and which (would) expedite the adoption of amendments" For this purpose it was agreed that a new procedure should be adopted for amending international treaties which established international regulations and standards. The new procedure would be based on the "tacit" or "passive" acceptance of amendments by governments. As was explained at the time, the procedure would mean that "the body which adopts the amendment at the same time fixes a specific time within which the Contracting parties will have the opportunity to notify either their acceptance of their rejection of the amendment, or to remain

---

[71] An account of the history of the adoption of the "tacit acceptance" procedure in IMO is given in a document of the Secretariat at the IMO website: <www.imo.org/Conventions/mainframe.asp>.

[72] Resolution A. 249(VII).

silent on the subject. In case of silence the amendment is considered to have been accepted by the party..."[73]

The tacit acceptance procedure has now been incorporated into the majority of the technical conventions of IMO, including those for the prevention of marine pollution. The effectiveness of this procedure has been demonstrated by the fact that IMO has been able to adopt and bring into effect amendments to several important provisions of international conventions within a time frame that would have been unimaginable under the traditional procedure which required express acceptance by a specified number (or stipulated proportion) of the States parties before the amendment could enter into force. Thus many important amendments to the Annexes of MARPOL and SOLAS as well as the London Dumping Convention have entered into force in two years or less following their adoption. For example, the amendments to Annex 1 of MARPOL, introducing a new global time-table for phasing out single-hull oil tankers, was adopted on 27 April 2001 and entered into force on 1 September 2002, i.e. less than eighteen months after the adoption of the amendment. Similarly, the 1993 amendments to the London Dumping Convention that were adopted in November 1993 entered into force on 20 February 1994. These amendments banned the dumping into the sea of industrial wastes by 31 December 1995 and prohibited the incineration at sea of industrial wastes. In all these cases, amendments that were deemed to be urgent were brought into operation by means of the tacit acceptance procedure, with very little delay.

(e) Another significant contribution by IMO to the development of international law in the field of environmental protection has been the use of liability and compensation as one of the tools for the prevention of marine pollution. It is now generally accepted that imposing strict liability on shipowners for pollution damage resulting from the operation of their ships can help to prevent pollution incidents because it tends to concentrate their minds on the need to take appropriate measures to avoid accidents to their ships[74] In addition, some of the provisions in the IMO conventions on liability and compensation provide useful incentives to the owners or operators of ships to take necessary measures to prevent or minimize pollution even after an accident has occurred. For example, the 1992 Civil Liability Convention encourages the owner or operator of a tanker that is involved in an accident to take preventive measures to avoid or reduce pollution damage from the accident. This is the effect of paragraph 6(b) of article I of the Convention which provides that the costs of preventive measures are entitled to compen-

[73] Note 73 supra.

[74] "it is basically true that international liability regimes for protection of the environment are also seen as an incentive to behave in a way that is environmentally sound. They thus fulfil a preventive function" See *Rüdiger Wolfrum et al.*, Environmental Protection by Means of International Liability Law (1999) page 2.

sation under the Convention[75]This provision is further amplified in article 4, paragraph 1, of the 1992 Fund Convention which expressly states that the "expenses reasonably made by the owner voluntarily to prevent or minimize pollution damage shall be treated as pollution damage for the purposes of (compensation)" The same approach is adopted in the 1989 Salvage Convention. For the first time, the Convention states that a salvor should be entitled to compensation for preventing pollution of the sea, even where the salvage operation did not succeed in saving the ship or the cargoes on board.[76] Under the traditional principle of "no cure no pay", a salvor would not be entitled to any reward if the salvage operation did not succeed in saving the ship or its cargo, even if it succeeded in preventing or reducing marine pollution from the incident.[77]

VII. Concluding Remarks

In the discharge of the responsibilities and functions entrusted to it by the IMO Convention and other international treaties, IMO has made a significant contribution to the international efforts to protect and preserve the marine environment, particularly from pollution arising from the operation of ships. It has provided a forum for the consideration and adoption of technical, legal and administrative regulations and standards for the prevention and control of marine pollution from ships. Using the mechanisms offered by its various organs and bodies, IMO has promoted the development of practical measures which can be taken by governments and private and public entities engaged in the shipping industry (a) to prevent accidents to ships at sea; (b) to ensure safe management and handling of cargoes that can cause pollution so as to prevent accidents or incidents that can lead to pollution of the marine environment; and (c) to prepare and equip themselves to be able to take necessary and appropriate measures to prevent, mitigate or minimize pollution when accidents occur. IMO has also played a valuable role in the implementation of the provisions of the United Nations Convention on the

---

75 Under Article I(6)(b), "pollution damage" (for which compensation is payable) is defined to include "the costs of preventive measures and further loss or damage caused by preventive measures"; and paragraph 7 of Article I defines "preventive measures" as "any reasonable measures taken by any person after an incident has occurred to prevent or minimize pollution damage".

76 Article 13 of the 1989 Salvage Convention provides that in determining the remuneration of the salvor, account should be taken of, *inter alia*, the skill and efforts of the salvors in preventing or minimizing damage to the environment". Further Article 14 establishes a "safety net" to provide to the salvor a special compensation equivalent to his expenses where the salvor has undertaken operations to assist a vessel that threatened damage to the marine environment but had not earned a reward under Article 13.

77 On the significance of the changes introduced by the 1989 Convention see, *Darling/Smith*, LOF 90 and the New Salvage Convention (1991) Chapter 2-3.

Law of the Sea. Finally, it has contributed in various ways to the elucidation and development of international law on the protection and preservation of the marine environment.

The Contribution of the European Union to Marine Pollution Prevention

Ludwig Krämer

I. Introduction

In public international law the European Union (EU) is called an organisation for *regional* economic integration, however, its environmental duty goes beyond economic integration, and is global in scope. The EU has fixed itself on the objective of "promoting measures at international level to deal with regional or worldwide environmental problems."[1] It has ratified all of the main UN environmental conventions, in particular, the Convention on the Law of the Sea.[2] However, it is widely considered that EU foreign policy is at present largely rhetoric. The lack of sincere policy also affects foreign environmental issues and protection of the marine environment. Indeed, the EC has not developed a consistent policy to address the protection of the marine environment, either at the global or regional level.

The first section of this contribution will describe the general policy approach of the EC with regard to maritime issues (II). Second, it will present and discuss EC measures taken over the last thirty years to prevent oceanic pollution (III). A third section will consider, in light of recent developments, whether significant changes to present policy are likely (IV). The final section will draw some conclusions (V).

II. Past EC Attempts to Establish a Policy on the Marine Environment

The European Union does not have an overall competence, but it may take action within the limits of the powers conferred upon it and the objectives assigned to it by the EC Treaty.[3] While the EC Treaty and the Treaty on European Union give the EU competence to act, apart from the area of environmental protection, in the

---

1 Article 174(1) EC Treaty.

2 Decision 98/392, OJ 1998, L 179, p. 1.

3 Article 5 EC Treaty.

areas of sea transport, fisheries, agriculture, energy and trade, as well as in foreign and development policy, it does not identify the oceans as a specific area of political activity. Thus, Community action with regard to oceans was never conceived or made operational in a coherent and consistent manner, nor did it integrate the different aspects of protection of the seas.

In the original EEC Treaty of 1958, the chapter on *Common Transport Policy* differentiated between inland waterway transport and sea transport, because in the 1950s most water transport services were supplied on inland waters. It was only in 1974 that the Court of Justice declared that the rules on transport policy also applied to sea transport.[4] A Treaty amendment of 1987 eliminated the substantive differences almost entirely and introduced the Commission's right to make legislative proposals on sea transport.

However, Community transport policy in general either avoided taking action where international transport conventions had already been adopted, or adhered to such conventions without explicitly transposing the provisions into Community secondary legislation. This policy choice was accompanied by another basic decision regarding the enforcement of legislation. Article 300(7) of the EC Treaty expressly provides that international conventions concluded by the EC "shall be binding on the institutions of the Community and on the Member States." Article 211 of the EC Treaty mandates the Commission to "ensure that the provisions of this Treaty and the measures taken by the institutions pursuant thereto are applied." However, the Commission never tried to enforce provisions of an international convention concluded by the Community against Member States who did not apply the convention, unless corresponding secondary legislation[5] was in place. Furthermore, Member States refused to let the EC undertake steps to initiate the procedure to become a member of the International Maritime Organisation (IMO). The result of this policy is Member States, and not the EC, generally led international initiatives and negotiations on maritime transport. Since transport policy was only concerned with limited aspects of maritime policy, initiatives incorporating a comprehensive approach to maritime issues were not taken in this sector.

Similar observations can be made with regard to the EC's *Common Fisheries Policy*. Based on Article 32ss of the EC Treaty, this policy aimed at optimising the amount of fish taken within Community waters and beyond. The interests of the fishing sector primed. Over the last fifteen years, it had become evident that over-fishing, pollution, and other activities had led to a strong reduction of fish availability globally, and in Community waters specifically. It was only at this point that the official orientation of the Common Fisheries Policy changed and

---

4 ECJ 4 April 1974, case 167/73, *Commission v. France*, ECR 1974, 359.

5 It was at the end of 2004 that the Court ruled, likely for the first time, and in an environmental case, that the provisions of in international convention concluded by the EC could be enforced against a Member State also in the absence of secondary Community law provisions, see ECJ 7 October 2004, case C-239/03, *Commission v. France*.

began looking into more general aspects of the sea and its natural resources.[6] However, the objectives of the Common Fisheries Policy laid down in Article 33 of the EC Treaty were not changed. The Community legislature and policy-makers at the oceans were once again concerned with the interests of fishermen and the fishing industry, and not with the state of the oceans in general. Policy relating to the areas of agriculture, energy, trade, and general foreign and development policy relating to maritime and ocean issues were handled accessorily and never included in any comprehensive matter.

When the EC started developing an *environmental policy* in the early 1970s, it attempted to address the issue of protection of the marine environment. This undertaking remained limited to the regional seas. The first Commission communication to the Council of 1971 suggested specific actions to combat pollution of the Rhine River, the Mediterranean, and the North Sea.[7] The Commission's approach was based on the belief that the majority of marine pollution stemmed from land-based sources.

The first proposal for an EC environment action programme[8] was much more specific. The Commission suggested

1. Setting up a European Agency to clean up the Rhine river basin
2. Legislative measures to fight the contamination of the oceans bordering the EC, in particular, the dumping of waste at sea and the discharge of municipal and industrial waste waters into the sea

However, these proposals proved to be too ambitious. France was unwilling to accept that the EC was entitled to deal with environmental issues under the EEC Treaty framework, arguing the environment was not mentioned as a subject of Community activity in the Treaty. This position, which contradicted other public statements by France,[9] resulted in the adoption of a compromise environment action programme in the form of a declaration "by the Council and by the Representatives of the Governments of the Member States meeting in the Council."[10] Several Member States objected to the establishment of a European body for the Rhine River, arguing this would duplicate the work of the Inter-

[6] In particular, see Regulation 2371/2002 on the conservation and sustainable exploitation of fisheries resources under the Common Fisheries Policy, OJ 2002 L 358, p. 59.

[7] Commission, SEC(71) 2616 of 22 July 1971, section 4.3. At that time, United Kingdom and Denmark were not yet members of the EC.

[8] Commission, Proposal for a Community environment action programme, OJ 1972 C 52, p. 1.

[9] It had been France which hosted and invited participants to the first ever European summit of Heads of State and Governments (Paris, October 1972) that united the then nine Member States and in its final declaration, largely inspired by France and agreed upon by all, invited the EEC institutions to launch an EEC environmental policy and an environment action programme.

[10] See First Environment Action Programme, OJ 1973 C 112, p. 5.

national Rhine Commission.[11] The announcement that legislation on waste dumping was to be elaborated upon was also met with resistance, on the ground that international conventions had already addressed this subject.

Underlying these objections was a controversy, which is best expressed in Dutch academic discussions: were the Rhine River and European oceans to be cleaned up by integrated action under EC auspices or by international cooperation of (EC Member) States? Under the direction of foreign department administrations, which easily prevailed over environmental concerns, it was concluded that international cooperation was preferable to integrated EC management.

The consequences were evident: the European Rhine Agency was never instituted and Commission proposals for directives on the dumping of waste at sea[12] were not discussed in Council. Attempts by the Commission to let the Community adhere to the MARPOL Convention, the Oslo Convention 1972 for the prevention of marine pollution by dumping from ships and aircraft, and the 1974 Helsinki Convention on the protection of the Baltic Sea, all failed because Member States vetoed such initiatives (e.g., Denmark with regard to the Oslo Convention, and Sweden with regard to the Helsinki Convention). It is not clear why the EC was allowed to conclude the 1974 Paris Convention for the protection of marine pollution from land-based sources and the 1976 Barcelona Convention for the protection of the Mediterranean Sea against pollution. It is supposed that Southern EC Member States had higher hopes for financial support from the EC with respect to Mediterranean matters, and that general political resistance against European integration, at least in maritime matters, prevailed in Denmark, United Kingdom, Netherlands, and Norway.

Such early decisions favouring intergovernmental cooperation in maritime environmental matters determined EC environmental policy over the next 25 years. Occasional accidents, such as on an oil rig in the Norwegian North Sea in 1977 or the Amoco Cadiz tanker off the French coast in 1978, led to some EC efforts to combat marine oil pollution, although these were largely unsuccessful. The EC did not try to establish an EC maritime policy on the environment, accepting the view that Member States should handle protection of the marine environment either individually or through participation in different international conventions. Also, the EC did not initiate activities within the framework of international conventions it had adhered to, but largely limited itself to follow initiatives undertaken by other contracting parties. In other words, ocean pollution prevention and clean up was left to Member States to deal with, or occurred as a result of international cooperation, but was not a subject of integrated EC action.

The difference in law between international cooperation and European integration is enormous and cannot be fully addressed here. However, some relevant differences should be mentioned:

---

11 This International Commission for the protection of the Rhine against pollution had been set up by virtue of an Agreement among riparian States in 1963.

12 Commission, proposal for a directive on the dumping of waste at sea, OJ 1976 C 40, p. 3; OJ 1985 C 245, p. 23.

1. Environmental decisions in international forums are almost always made unanimously, giving negotiating or contracting parties a *de facto* right to veto decisions. Environmental decisions within the EC are taken by majority decisions.
2. Under international law, each country may introduce proposals for new measures. Countries are likely to make proposals that are in their own best interest and which do not necessarily benefit other countries, the oceanic environment in question, or the environment in general. Under EC environmental law, the Commission has the exclusive right to propose EC measures. The Commission is obliged to act in the general interest of the EC, not of a specific country.
3. Under international law, governments assent to the final legal measure (e.g., convention, agreement, etc.). Under EC environmental law, the European Parliament and the Council (which is composed of governments) must jointly agree to the final measure. The European Parliament is a directly elected body. Furthermore, before a text is adopted regarding environmental matters the European Economic and Social Committee and the Committee of the Regions must be consulted.
4. In international law, any agreed text must pass the scrutiny of national ratification before it becomes effective at the national level. In EC law, agreed to texts are binding, and Member States that do not comply with these texts infringe their obligations.
5. In international law, there is almost no implementation and enforcement process. The main source of information for convention secretariats are reports provided by contracting States who have little inclination to report on their omissions and failures to comply with their international legal obligations. States that do not honour their environmental commitments are not exposed to sanctions. There is practically no court to settle disputes regarding environmental matters. Existing mechanisms deal with disputes between States, not instances where the action of a State may harm the environment. In EC law, a relatively sophisticated implementation and enforcement mechanism exists which may lead to condemnations by the EC Court of Justice,[13] and has the effect that eventually all Member States rally to respect the legislation. Environmental court litigation is predominately initiated by the Commission. This ensures the rule of law is respected and the environment is protected, both general interests held by the EC.
6. Elaboration and implementation of international environmental law is the matter of specialists (i.e., diplomats and quasi-diplomats) who normally act outside public awareness and attention. This is of particular importance, as the environment is an interest with no voice of its own. The development

[13] In environmental matters alone, as of mid-2006, the ECJ has given about 500 judgments against Member States. Interestingly enough, there is not a single environmental dispute between Member States which was ever decided by the Court, though Article 227 EC Treaty provides for an action of one Member State against the other.

and implementation of EC environmental law is extensively discussed within the European Parliament by stakeholders, journalists, environmental groups, and other interested bodies, both at the European and national levels.

III. EC Measures to Prevent Marine Pollution

Pollution prevention measures also affect the marine environment. For example, measures to reduce air emissions from industrial installations will reduce the input of pollutants into seas from the air. The following enumeration is limited to EC measures having the greatest direct impact on marine pollution prevention. While prevention is a relevant aspect of any environmental policy, a wide range of opinion exists regarding how far preventive measures should go. For that reason, it is much easier to adopt measures that reduce or limit pollution, which will also inevitably have a preventive effect.

The EC has never published a comprehensive paper on the state of the marine environment, the measures taken or planned respecting the marine environment, or any other information on the marine environment as such. For example, the first EC report after the fall of the Berlin wall and also the first report on the state of the European environment elaborated by the European Environment Agency, appeared in 1995.[14] It contained a chapter entitled "The assessment, the seas" where a number of problems were generally described. The section on "Problems" only listed a chapter on "coastal zone threats and management". The 1999 report[15] by the same agency described (fresh) "water stress" and "coastal and marine zones." The third report[16] discussed "water", however, again mainly fresh water aspects. The fourth report of 2005 had seven sections on water, none of them on oceans. In three of these sections marine waters are at least mentioned, namely, the sections on nutrients and chlorophylls "in transitional, coastal and marine waters" and the section on bathing waters.[17] This low attention to ocean and sea questions illustrates the strong priority the EU gave to land-based activities.

Discharge of dangerous substances. In the wake of the international conventions of Oslo, Paris, Helsinki, and Barcelona to protect European seas from pollution, it was almost natural that the EC should also try to contribute to marine environmental protection. The first measure introduced with regard to ocean pollution was Directive 76/464 on the discharge of certain dangerous substances into Commu-

[14] European Environment Agency, Europe's Environment. The Dobris Assessment. Luxembourg 1995.

[15] European Environment Agency, Environment in the European Union at the turn of the century. Luxembourg 1999.

[16] European Environment Agency, Europe's environment: the third assessment. Luxembourg 2003.

[17] European Environment Agency, The European Environment. State and outlook 2005 (Copenhagen, 2005).

nity waters.[18] The Directive followed the example of different international conventions and established two lists of substances and groups of substances. The first list (List I) concerned substances that were toxic, persistent, or bioaccumulative. For substances on List I, emission limit values for discharges were to be fixed at the EC level, which could not be exceeded by the individual permits. However, EC directives at that time had to have unanimous agreement.[19] The United Kingdom strongly opposed emission limit values, arguing that its short, quick-flowing rivers would quickly carry away and dissolve pollutants, and this natural advantage should not be lost with the fixation of uniform emission standards. All of the other nine Member States favoured uniform EC emission standards. As no consensus could be reached, the Directive provided for uniform emission standards for List I substances and equivalent quality standards. Member States were free to decide which option they would choose.

A second list (List II) grouped other dangerous substances. For these substances quality standards would be fixed by EC legislation. Member States were obliged to adopt clean up programmes in order to reduce water pollution resulting from List II substances. The clean up programmes also had to contain quality standards. List I substances, for which the EC had not yet established emission limit values, were to be treated as List II substances.

Directive 76/464 had a disappointing implementation. Member States were able to agree to 129 priority substances to place on List I, for which EC emission limit values and quality objectives were to be fixed. However, work on the specific substances was held up, slowed, and finally stopped. Only 17 substances were regulated between 1976 and 1991. Almost none of the Member States adopted programmes respecting List II substances, and most of the other provisions in the Directive were neglected by Member States and the Commission itself. Despite numerous public resolutions and statements by the Council calling for an acceleration of the work on List I substances, no change was observed. In 1991 efforts toward putting this Directive into operation stopped without any public statement, declaration or justification; an obvious silent cooperation between the Commission and Member States which has heretofore not been explained. The Commission started numerous infringement proceedings against Member States, and an impressive number of Court judgments found that Directive 76/464 had been infringed. However, such action did not result in significant changes to the situation.

Under the influence of strong pressure, particularly from the United Kingdom, and rising opposition to an EC environmental policy and actions, the Commission made proposals to reorient the EC water policy.[20] The new Commission proposal

[18] Directive 76/464 on pollution caused by certain dangerous substances discharged into the aquatic environment of the Community, OJ 1976 L 129, p. 23.

[19] Majority decisions for the environment were introduced as of November 1993 (Maastricht Treaty).

[20] See Commission proposal for a directive on the ecological quality of water, OJ 1994 C 222, p. 6; see also Commission, Communication on the European Community Water Policy, COM (96) 59 of 21 February 1996.

on a water framework directive on water policy[21] proposed abandoning EC emission limit values, and instead concentrating on water quality provisions. Member States welcomed this approach, as the experience of 1976 demonstrated that Member States' respect of EC quality objectives could not be controlled by the EC Commission. The European Parliament did not agree with this concerted policy of deregulation. It insisted that Member States had committed themselves at several North Sea Conferences, in Council resolutions, and in other public statements to reduce water pollution, and that these commitments required more than just the orientation towards nonbinding quality objectives fixed by Member States.

After intense debate, the Council accepted Parliament's point of view. The new Directive 2000/60[22] provided in Article 16 for EC measures combating water pollution by individual pollutants, and for the stoppage or phasing out of discharges, emissions and losses of priority hazardous substances within 20 years. A list of priority substances was established in 2001 that enumerated 33 substances or groups of substances from which 25 were identified as priority hazardous substances.[23] In 2006, the Commission proposed a directive that fixes quality objectives for dangerous substances, and would repeal by 2012 existing emission limit values. Member States are to take the measures by providing river basin management plans.[24] This proposal represents an about-face by the EC with respect to fixing quality objectives, although it has been proven in the past that such quality objectives cannot be seriously enforced and controlled. It remains to be seen, how the Council and the European Parliament will react to the proposal.

Overall, the EC's action on hazardous substances is disappointing. Within thirty years, the EC has managed to set up priority lists for hazardous substances, but has not significantly reduced the discharge of hazardous substances into EC waters. This conclusion is nuanced with regard to the 17 substances regulated under Directive 76/464. However, even for the 17 regulated substances, which were not necessarily the most frequently used substances, the provisions were relatively vague, contained considerable exceptions and derogations, and allowed Member States and polluters to escape compliance requirements for fixed quality standards. Despite plans to reduce hazardous substances by means of priority lists and other attempts to establish a strategy for joint activities in this area, little coordination among Member States has taken place. The different EC provisions allowed Member States that were determined to reduce marine pollution to do so, but were not stringent enough or sufficiently monitored to impose compliance when Member States were unwilling to reduce pollution.

---

21 Commission, proposal for a directive establishing a framework for Community action in the field of water policy, OJ 1997 C 184, p. 20.

22 Directive 2000/60 establishing a framework for Community action in the field of water policy, OJ 2000 L 327, p. 1.

23 Decision 2455/2001establishing the list of priority substances in the field of water policy and amending Directive 2000/60, OJ 2001 L 331, p. 1.

24 Commission, proposal for a directive on environmental quality standards in the field of water policy, COM (2006) 397 of 17 July, 2006.

Quality of bathing waters. Another piece of important legislation that warrants mention is Directive 76/160 on the quality of bathing waters.[25] Directive 76/160 required the monitoring of coastal and inland bathing waters with regard to ten different pollutants, in particular, faecal coliforms. During bathing seasons, pollutant parameters needed to be regularly controlled, and pollution clean-up measures had to be undertaken or bathing prohibitions fixed. The Directive will be repealed as of 2014 and replaced by one that reduces the controlled parameters to two and introduces more flexibility for local or national authorities regarding water quality monitoring.[26]

At present, there are 14 000 coastal waters which come under the provisions of Directive 76/160. The rate of compliance with the Directive on bathing waters was slow, but progressively improved over the years because many local authorities saw an opportunity to advertise clean bathing water to attract tourists. The principal measures that improved the quality of bathing waters related to waste water treatment from urban agglomerations and industrial outlets.

Oils. Directive 75/439 on waste oils[27] prohibited the discharge of waste oil into coastal and surface waters. However, it only applied to lubricating oil and not to all oils. Several sea incidents involving hydrocarbons, such as an oil platform incident in the North Sea in 1977 and the Amoco Cadiz accident in 1978, provoked Commission communications to the Council asking for stronger measures to be taken against oil pollution of the sea. Such measures were strongly favoured by France. The Council managed to agree to a resolution, which asked for the elaboration of an action programme to reduce oil pollution at sea, and agreed to an exchange of information on ocean oil pollution.[28] However, all further proposals by the Commission concerning the establishment of common structures, emergency plans, and demonstration programmes were rejected. A consultative committee for sea pollution by hydrocarbons established in 1980[29] never became significant at the EC level.

Activities of the offshore oil industry were never the subject of any Community measure.

Ship safety. The improvement of ship safety, particularly respecting tankers transporting hydrocarbons, is closely linked to the problem of oceanic oil pollution. Community measures concentrated on the prevention of accidents by ships and to the increase of transport safety. Internationally, the International Maritime Organisation (IMO) is charged to deal with such issues. Once more, the EC was not able to become member of IMO, but it participates as observer in IMO meetings. EC policy on ship safety consisted of leaving the IMO to take initiatives

25 Directive 76/160 on the quality of bathing waters, OJ 1976 L 31, p. 1
26 Directive 2006/7 on the quality of bathing water, OJ 2006 L 64, p. 37.
27 Directive 75/439, OJ 1975 L 194, p. 1.
28 Resolution of 26 June 1978, OJ 1978 C 162, p. 4; Decision 81/971, OJ 1981 L 355, p. 24 and Decision 86/85 (exchange of information) OJ 1986 L 77, p. 33.
29 Decision 80/686, OJ 1980 L 188, p. 11.

and then to recommend that Member States accede to the different IMO agreements.

In early 1993, an accident involving the oil tanker "Braer" caused considerable oil pollution off the coast of Shetland. Shortly after this disaster, the Commission adopted a communication on a common policy on safe seas, which the Council welcomed in a resolution.[30] The *Braer* accident caused the United Kingdom and other EC Member States (e.g., Netherlands, Denmark) to change their policy and to accept a more active role for the EC with respect to ship safety, which would be steered under transport, and not environmental auspices. Since early 1993, a considerable amount of EC legislation was adopted and regularly updated, which had at least an indirect effect of reducing and preventing further sea pollution. This legislation set out minimum requirements for vessels carrying dangerous or polluting goods bound for or leaving Community ports,[31] established common rules and standards for ship inspection, survey organisations, and relevant activities of maritime administrative bodies,[32] regulated the minimum level of training of seafarers,[33] instituted a regulation on tonnage measurement of ballast spaces in segregated ballast oil tankers,[34] a directive concerning the enforcement of international standards for ship safety, pollution prevention and shipboard living and working conditions (port State control)[35] and directives on safety at fishing vessels,[36] passenger ships[37] and ferries[38]. Accidents involving the ship "Erika" off the coast of France in 1999 and the tanker "Prestige" off the coast of Spain in 2003 led to further legislative activities. A European Maritime Safety Agency and a Committee on Safe Seas (COSS) were created,[39] ship inspection

30 Commission, COM (93) 66 of 24 February 1993; Council Resolution of 8 June 1993, OJ 1993 C 271, p. 1; see also European Parliament, OJ 1993 C 42,, p. 16; OJ 1994 C 91, p. 301.

31 Directive 93/75, OJ 1993 L 247, p. 19.This Directive was later replaced by Directive 2002/89 establishing a Community vessel traffic monitoring and information system, OJ 2002 L 208, p. 10.

32 Directive 94/47, OJ 1994 L 319, p. 20.

33 Directive 94/58, OJ 1994 L 319, p. 28. This Directive was replaced by Directive 2001/25 on the minimum level of training of seafarers, OJ 2001 L 136, p. 17. One should imagine an EC directive on the minimum level of training fort truck drivers, economic managers or other persons who might exercise an environmentally sensitive activity, to realise the shift in policy which the 1993 events have caused.

34 Regulation 2978/94, OJ 1994 L 319, p. 1.

35 Directive 95/21, OJ 1995 L 157, p. 1.

36 Directive 97/70 setting up a harmonised safety regime for fishing vessels of 24 m length and over, OJ 1997 L 34, p. 1.

37 Directive 98/18 on safety rules and standards for passenger ships, OJ 1998 L 144, p. 1.

38 Directive 1999/35 on a system of mandatory surveys fort he safe operation of regular ro-ro ferry and high-speed passenger craft services, OJ 1999 L 238, p. 1.

39 Regulation 1406/2002 establishing a European Maritime Safety Agency, OJ 2002, L 208, p. 1; Regulation 2099/2002 establishing a Committee on Safe Seas and the prevention of pollution from ships (COSS), OJ 2002 L 324, p. 1.

procedures harmonised,[40] the safety provisions for bulk carriers[41] and ro-ro passenger ships[42] and general ship and port safety measures strengthened.[43] Of particular importance were the measures to provide for double hull or single hull tankers.[44] Other decisions worthy of mention include the decision to create port reception facilities for waste from ships,[45] and to oblige ships using EC ports to pay for these facilities regardless of whether they actually use them.

Industrial activities. A Commission proposal to reduce water pollution by the paper pulp industry[46] was not adopted by the Council. Following an incident in the Mediterranean that resulted in environmental pollution, the EC adopted a directive on waste from the titanium dioxide industry that provided strict conditions for such discharges in waters in 1978[47]. The Directive, heavily opposed by industry, was later completed with provisions on clean-up programmes and further discharge restrictions.[48] Member States reluctantly implemented these provisions. They eventually became unimportant when new applications allowed for the recycling of waste from the titanium dioxide industry and its use in agriculture, to the point that the European Commission has considered proposing their repeal.

By the end of the 1970s, the Commission abandoned its policy to regulate individual polluting industries and has refused to reconsider its reimplementation.

In general, EC measures on industrial installation were largely limited to air emissions. However, a 1984 directive on air emissions from industrial installations[49] was replaced in 1996 by a directive on integrated pollution prevention and control from industrial installation.[50] The new Directive required the use of the best available techniques to reduce emissions into the air, water and soil, for all new installations and by 2007 for existing installations. The limited practical impact of the 1996 Directive was due to the loosely defined notion of best

40 Regulation 884/2005 laying down procedures for conducting Commission inspections in the field of maritime security, OJ 2005 L 148, p. 25.

41 Directive 2001/96 establishing harmonised requirements and procedures fort he safe loading and unloading of bulk carriers, OJ 2001 L 13, p. 9.

42 Directive 2003/25 on specific stability requirements for ro-ro passenger ships, OJ 2003 L 123, p. 22.

43 Regulation 725/2004 on enhancing ship and port facility security, OJ 2004 L 129, p. 6.

44 Regulation 417/2002 on the accelerated phase-in of double hull or equivalent design requirements for single oil tankers, OJ 2002, L 64, p. 1. Following the Prestige accident, a further acceleration was decided by Regulation 1726/2003, OJ 2003 L 249, p. 1.

45 Directive 2000/59 on port reception facilities for ship-generated waste and cargo residues, OJ 2000 L 332, p. 81.

46 OJ 1975 C 99, p. 21.

47 Directive 78/176, OJ 1978 L 54, p. 19.

48 Directives 82/883, OJ 1982 L 378, p. 1, and 92/112, OJ 1992 L 409, p. 11.

49 Directive 84/360 on the combating of air pollution from industrial plants, OJ 1984 L 188, p. 20.

50 Directive 96/61 on integrated prevention and pollution control, OJ 1996 L 257, p. 26.

available techniques and that it would only apply to existing installations as of 2007. A Directive on waste incinerators[51] also dealt with water discharges from these installations, although most incinerators are not placed at the coasts.

Wastewater. In 1991, the EC adopted a directive on urban wastewater.[52] This Directive provided that by 2005 all local authorities with more than 2 000 persons had to be equipped with canalisation and with secondary wastewater treatment installations. Although this timetable has not been completely respected,[53] it has led to considerable investments in all Member States which have contributed to reductions in coastal water pollution. Overall investments were estimated to be more than 149 billion € to build some 40 000 waste water treatments plants. Pressure from the European Commission and the threat of legal action will eventually lead complete compliance with the Directive's requirements.

Agricultural pollution. In 1991 the EC also adopted a directive on the protection of waters from nitrate pollution caused by agricultural sources.[54] The Directive limited the amount of nitrate in waters, including coastal waters, to 50 mg per litre and requested the taking of measures, including the reduction of livestock in areas where nitrate limits were at risk of being exceeded. Implementation of this Directive appeared difficult, as a result of strong resistance from farmers, in particular of intensive agriculture installations, and from governments not inclined to risk confrontations with the farming lobby.[55]

Other aspects. As previously mentioned, the EC did not manage to adopt a directive on the dumping of waste at sea. It also omitted to deal with the discharge of radionuclides into the sea, leaving this matter to be handled by international organisations or Member States. Concerns such as litter and noise were also not addressed. While the EC adopted provisions on natural habitats and endangered fauna and flora species,[56] it was not able to take concrete measures to stop the by-catch of cetaceans in fisheries, or obtain the designation of protected habitats for seals or marine mammals in general until 2006. Finally, Community legislation on environmental liability, which was adopted in 2004,[57] exempted the largest environmental disasters in Western Europe from its field of application by letting the international conventions on liability for oil pollution prevail over the EC provisions.

---

51 Directive 2000/76 on the incineration of waste, OJ 2000 L 332, p. 91.

52 Directive 91/271, OJ L 135, p. 40.

53 See the informative reports on the implementation of Directive 91/271, COM(98) 775 of 15 January 1999 and COM(2004) 248 of 23 April 2004.

54 Directive 91/676 OJ 1991, L 375, p. 1.

55 See the implementation report, Commission, Implementation of Directive 91/676. (2002) Luxembourg.

56 Directive 92/43 on the conservation of natural habitats and of wild fauna and flora, OJ 1992 L 206, p. 7.

57 Directive 2004/35 on environmental liability with a view of preventing or remedying environmental damage, OJ 2004, L 143, p. 56.

Overall, the EC contribution to the prevention of marine pollution is limited. From an environmental policy perspective, the EC took a number of measures to improve the condition of the marine environment and prevent further impairment. Also successful were EC measures taken in relation to the safety of transport at sea, in particular those following the Braer, Erika and Prestige accidents. It is notable that measures aimed directly at improving the marine environment are lacking; such protection is a welcome side effect of measures that are principally aimed at reaching other objectives.

A number of factors may help to explain this limitation:

1. Data on threats to the oceans, degrees of pollution, and endangered species, such as marine mammals, are scarce. Such data are usually made available by Member States themselves, which have an interest in collecting and publishing data favourable to their fisheries, transport, tourism, or other policies. For example, Mediterranean countries have not shown much interest in financing research that might require dramatic reductions in fishery activities or asking for measures that would affect the tourist sector. Data availability is better with regard to the Baltic Sea and the North Sea, but such data mainly feed systems established under the Helsinki and OSPAR Conventions. The European Environment Agency has made little data available on the marine environment that would allow for general discussion on the state of the environment and the necessity to prioritise actions.

   A good example of this problem is cadmium. Cadmium was identified in 1976 as a substance requiring priority action with regard to marine discharges and its overall emissions into the environment.[58] A discharge directive of 1983[59] was disappointingly vague, and the practical application of the Directive was hardly monitored by the Commission and Member States. Since the early 1990s the Directive had almost fallen into oblivion, because no data were published. Cadmium was identified in 1987 as particularly threatening, and a Council Resolution recommended substituting it wherever technically possible.[60] The Council's suggestion led to some restrictions on its use, but the directive was full of derogations, transition periods, and other provisos.[61] In 1997, cadmium was identified as a priority substance for a risk assessment,[62] an indication that the political decision of 1987 to substitute it wherever possible did not exist. This risk assessment is, by mid-2006, not yet finished. Cadmium was also placed on a 2001 list as a substance that was

---

58 See Directive 76/464 (Note 18 supra), annexe I. See also already first environment action programme, OJ 1973, C 112,, p. 13; Council Resolution of 24 June 1975, OJ 1975 C 168, p. 4.

59 Directive 83/513 on limit values and quality objectives for cadmium discharges, OJ 1983 L 291, p. 1.

60 Council Resolution of 25 January 1988, OJ 1988 C 30, p. 1.

61 Directive 91/338, OJ 1991 L 186, p. 59. This Directive was to be reviewed at regular intervals, which has not yet occurred.

62 Commission Regulation 143/97, OJ 1997 L 25, p. 13.

particularly hazardous to waters,[63] without any legislative or other consequences until today. For practically 30 years, the EC legislature has been going in circles and simply not making decisions, often arguing that sufficient data are unavailable. Cadmium is a well-researched pollutant, and the EC has taken little action in preventing its discharge into waters.

2. Member States declared a preference that regionalised marine protection conventions handle marine issues. This tendency is particularly true of Member States bordering the North Sea and the Baltic Sea (e.g., Denmark, Netherlands, United Kingdom, Germany). Member States that border the Mediterranean Sea are less likely to prefer a regionalised treatment of data and a corresponding fixing of priorities. However, Mediterranean countries do not often cite the relevant data. As a consequence, the Barcelona Convention is not active, and overall, little action has been taken to protect the environment. The protection of the Black Sea marine environment is not yet a subject of attention of the EC legislature.
3. Historically, the evolution of legislation within the EC shows that administrations of Member States learn through accidents. Increased legislative activity at European level resulted from oil spills involving the Braer, Brent Spar, Erika, and Prestige. However, accidental pollution of seas is not the principal problem. Degradation of the marine environment is mainly slow, but progressive, and results from discharges, eutrophication, increased transport, pollution from shipping operations and offshore activities, and over-fishing. The EC needs to address these causes using preventive measures such as planning and programming, strict monitoring, complete transparency, and public participation and discussion. It is doubtful at this stage whether Member States are really prepared to have an integrated, transparent, and fully monitored EC marine policy; or whether they prefer the present state of affairs, where priorities are discussed outside of public awareness, measures are decided by consensus, outside enforcement is lacking and there is no real disadvantage to not reaching targets.

IV. An EC Strategy for the Protection of the Marine Environment

In view of the EC's past record in addressing specific marine environmental issues, it was surprising when the European Commission announced a new initiative in January 2001 to develop a "strategy for marine protection."[64] The Commission stated that knowledge about the structure and functioning of the marine environment remained surprisingly limited, and that the understanding of the human impact on marine ecosystems was very weak; in reality the impact was great, in particular via pollution.

---

63 Decision 2455/2001(Note 22 supra).

64 Commission, Environment 2010, our future – our choice – Communication on the sixth environment action programme of the European Community, COM(2001) 31 of 24 January 2001.

> "Other pressures come from ships that empty their oil tanks, ship accidents and heavy human utilisation of coastal areas. The introduction of non-native (alloctone) species in new marine environments, can also give rise to environmental stress. This has led to an increasing disturbance and pollution of our seas with negative effects on marine habitats and life. There is a decline of fisheries reported in almost all regional seas. Many fish stocks are over-exploited. There is a need to reduce pressure from fishing... protection of the marine environment ...needs an integrated strategy."[65]

The announcement of a (thematic) strategy on the marine environment had not been preceded by any detailed analysis of marine problems. As with the other thematic strategies announced by the Commission – on soil protection, air pollution, biodegradable pesticides, sustainable use of resources and waste recycling – the marine environmental strategy was intended to make it less obvious that the Sixth Environment Action Programme did not have many constructive ideas regarding European environmental protection policies for the following ten years.

The Commission's proposal sailed through the European institutions without many problems, and surprisingly, questions of competence were never raised. The final decision on the Sixth Environment Action Programme, made jointly by the European Parliament and the Council[66] asked the EC to adopt "a thematic strategy for the protection and conservation of the marine environment taking into account, inter alia, the terms and implementation obligations of marine conventions, and the need to reduce emissions and impacts of sea transport and other sea and land-based activities."[67]

However, the European Parliament insisted that the thematic strategies under the Sixth Environment Action Programme should be adopted in 2005 by way of the co-decision procedure under Article 251 EC Treaty. The Parliament argued that under Article 175(3) of the EC Treaty it would have had co-decision power and could have co-decided on the content of the strategy for marine environment protection. Therefore, that the thematic strategy was to be agreed to with only some delays after the adoption of the Sixth Environment Action Programme should not deprive the European Parliament of this co-decision power.

In 2002, the Commission made a communication to the other EC institutions to prepare the thematic strategy66. The communication stated that it was premature to provide an integrated approach to marine environmental protection, and instead followed an action and sector oriented approach, which addressed the following issues:

1. Review of the current information and identification of the main threats
   This chapter reviewed, in general terms, the principal threats to the marine environment, including threats to biodiversity, threats from pollutants, eutro-

[65] Commission (Note 51 supra) 35. In its proposal for a decision on the sixth environment action programme, OJ 2001 C 154E, p. 218, Article 5(7), the Commission used the notion of "thematic strategy".

[66] Decision 1600/2002 laying down the Sixth Environment Action Programme, OJ 2002 L 242, p. 1.

[67] Decision 1600/2002 (Note 64 supra) Article 6(2)(g).

phication, human activity, etc. and briefly described EC actions in different sectors.

2. Development and implementation of policies to control these threats
   This chapter described all international activities undertaken by specialised organisations under regional conventions and the EC, sub-divided for hazardous substances, eutrophication, chronic oil pollution, radionuclides, health and environment issues and maritime transport.
3. Identification of gaps in knowledge in order to monitor assessment and research
   This chapter attempted to address the question of whether the information necessary to protect and conserve the marine environment was available, and then went on to identify several gaps in existing knowledge.
4. Drawing of operational conclusions of what needs to be done
5. This section concluded that the "overview of existing monitoring and assessment programmes, and the knowledge they have generated, reveals a significant number of information gaps on the state of the marine environment and on the effectiveness of the existing measures. Consequently, it is, in many cases, unclear whether and which additional protection measures should be considered as well as the administrative level at which they should be considered."
6. Identification of the operational and institutional EC objectives
   Under this heading, the Communication gave an outline of possible content for a marine strategy.
7. Setting out of an action plan to define and develop a thematic strategy by 2004

Finally, the Communication identified 14 objectives that the EC's marine strategy should attempt to meet:

1. Halt biodiversity decline by 2010[68]
2. Ensure a sustainable use of biodiversity through the protection and conservation of natural habitats and of wild fauna and flora
3. Change in fisheries management to reverse the decline in stocks and ensure sustainable fisheries and healthy ecosystems, both in the EU and globally
5. Reach concentrations of hazardous substances in the marine environment close to background values for naturally occurring substances and approaching zero for man-made synthetic substances
6. Eliminate human induced eutrophication problems by 2010
7. Reach concentrations in the marine environment close to background values for naturally occurring radioactive substances and approaching zero for artificial radioactive substances by 2020

68 The European Council of Heads of States and Governments already agreed to this objective at their Gothenburg Summit in 2001. It was repeated in Decision 1600/2002 on the Sixth Environment Action Programme of the EC (Note 64 supra).

8. Ensure compliance with existing discharge limits of oil from ships and offshore installations by 2010 at the latest, and eliminate all discharges from these sources by 2020
9. Eliminate marine litter arising from illegal disposal by 2010
10. Reduce the environmental impact of shipping
11. Achieve a high quality of the environment where levels of contaminants do not give rise to significant impacts on or risks to human health and wellbeing
12. Implement Community commitments made in the Kyoto Protocol
13. Realise more effective co-ordination and cooperation between the different institutions and regional and global conventions, commissions and agreements
14. Pursue this strategy at global level and implement international Conventions and codes of practice
15. Improve the knowledge base on which marine protection policy is based

To achieve these objectives, the Commission announced 23 actions which it would pursue during the next years; some of the actions were rather precise, and others vague.

On 4 March 2003, the Council adopted conclusions welcoming the Communication, endorsing its approach, and proclaiming it constituted a good basis for further development of Community action. It requested that the present a thematic strategy to protect and conserve the marine environment as soon as possible.

The Communication was extensively discussed with interested groups and persons. Following a large stakeholder conference in Koge, Denmark in December 2002, four working groups were set up to discuss key aspects of the Communication. Sixteen internal commissions and conventions were consulted, and a number of countries sharing the oceans and seas with the EU, as well as industry and civil society organisations, also participated. The results of the consultation process were discussed at a major conference in November 2004 in Rotterdam and an internet consultation was launched in spring 2005.

While discussion of this Communication and its content took place at the EC level, in 2004 the Barroso-Commission took office with a relatively clear political objective to deregulate, reduce environmental activities, and promote economic interests. This re-orientation of the Commission also impacted EC marine policy. Within the new Commission, the Malta Commissioner Borg obtained the responsibility for "fisheries and maritime affairs," without this responsibility being delimited with the environmental responsibility that was given to the Greek Commissioner Dimas.

In spring 2005, President Barroso and Commissioner Borg made a joint communication to the Commission[69] entitled: "Towards a future maritime policy for the Union: A European vision for the oceans and seas." The document discussed the case for a holistic approach to oceans and seas, the international context, the case for acting at European level, working towards an EU maritime

[69] As an internal paper, the document was neither dated nor numbered. It was presented in early March 2005.

policy, a green paper on a future EU maritime policy and the organisation of the maritime policy task force. Its main purpose appeared to be to announce that in the first half of 2006 the Commission would adopt a green paper on a future EU maritime policy, and establish a Commission Task Force to elaborate this green paper. The task force would be comprised of the Commissioners for enterprise and industry, transport, environment, fisheries and maritime affairs, research, and energy, and would be chaired by Commissioner Borg.

In substance, the Borg paper did not go beyond generalities, and limited its scope by indicating that the future green paper would constitute a first step towards an all-embracing EU maritime policy which would try to strike the right balance between the economic, social, security, safety, and environmental dimensions of sustainable development. The environmental thematic strategy was mentioned only once, and was described as providing valuable input to the future maritime policy.

It can only be speculated why President Barroso announced a green paper on maritime policy when there was already Council consensus to develop a framework directive, or another legally binding instrument as a thematic strategy for the marine environment. Since Article 6 EC Treaty compels environmental requirements be integrated into the elaboration and implementation of all other EC policies, it would have been easy to extend this thematic strategy, where necessary, to areas such as research, fisheries or energy. However, one might question what would be achieved by an all-embracing approach to maritime policy; while sectors such as fisheries and energy, or transport and industry, have some points in common, they also have distinct features which are not easily melded in a common paper. An integrated approach may ultimately constitute an attempt to impose too much of an environmental approach on maritime issues.

In October 2005, the Commission made a proposal for a directive "establishing a framework for Community action in the field of marine environmental policy marine strategy directive)."[70]

The explanatory memorandum of the proposal stated that past measures to control and reduce pressures and impacts on the marine environment had resulted "in a patchwork of policies, legislation, programmes and action plans at national, regional, EU and international level. The general picture that emerges from this policy framework is a mixed one. On the positive side, some progress has been made in certain areas, e.g. in reducing nutrient inputs or pollution from hazardous substances in particular heavy metals. However, overall, the state of the marine environment has been deteriorating significantly over recent decades. As a result, Europe's oceans and seas are under threat, in some cases to the extent that their structure and function is jeopardised."

The proposal sets a framework for the development of marine strategies "designed to achieve good environmental status in the marine environment (by the year 2021 at the latest)." It asked each Member State to develop a marine strategy for its European marine waters according to a plan of action. The plans had to provide for an initial assessment (Article 7), a determination of good environ-

70 Commission, COM (2005) of 24 October 2005.

mental status (Article 8), the establishment of a series of environmental targets (Article 9), the establishment of a monitoring programme (Article 10) and a programme of measures designed to achieve good environmental status (Article 12). The Commission had to be informed of all the measures taken by Member States and would have the possibility to reject the measures decided (Articles 11 and 15).

In order to establish good environmental status, Member States were to determine for each marine region a set of characteristics for which the proposal gave some guidelines. Sets of characteristics were then sent to the Commission, which had to lay down "generic qualitative descriptors, detailed criteria and standards for the recognition of good environmental status" (Article 8).

Member States with marine waters within the same marine region or sub-region, as defined in Article 3, were to co-ordinate their actions and "where practical and appropriate…use existing structures established in that Marine Region or Sub-Region" (Article 5(1)). Member States shall also as far as possible, build on existing programmes and activities developed in the framework of structures stemming from international agreements.

It is too early to assess whether, and through which processes, the EC institutions will approve this proposal, and what its final form will be. The following is clear:

1. The proposal largely mirrors Directive 2000/60, which mainly affects EC fresh waters. It does not intend to ensure the protection of the marine environment by EC measures. Rather, the measures are to be taken and implemented at regional level. While the overall objective of the proposal is to achieve a good environmental status, measures taken within each region may differ.
2. The proposal does not compete with regional marine environmental conventions. Rather, it tries to build on existing measures, plans, or programmes adopted under these conventions. Thus likely consequences will be that for the Baltic Sea and the North East Atlantic, existing measures might need some adaptations and alignments, but will essentially remain intact. In contrast, it is likely that Member States bordering the Mediterranean Sea, Spain, France, Italy, Slovenia, Greece, Malta, and Cyprus, will form several sub-regions that will need to initiate measures that comply with the requirements of the proposal.
3. The measures adopted for each region and sub-region will not be identical or even equivalent. Differences can be attributed to the imprecise definition of "good environmental status" in the proposal,[71] which nowhere states what is a *good* status, rather limiting itself to indicate what an environmental status is.

---

[71] See proposal, Article 1(2): "'environmental status' means the overall the state of the environment in marine waters taking into account the structure, function, and processes of the constituent marine ecosystems, together with natural physiographic, geographic and climatic factors, as well as physical and chemical conditions including those resulting from human activities in the area concerned".

4. It is clear that the two basic stones of this whole building are on one hand, the political determination to clean up the oceans and seas surrounding the EC, and on the other hand, the strong enforcement of the measures adopted to achieve the good environmental status by 2021 (or later). What is new about this approach is that the measures adopted in all of the regions and sub-regions would be EC measures. This means that the Commission has the legal obligation to ensure these measures are actually applied.[72] If necessary, infringement procedure set up under Article 226 EC Treaty is available. It is speculative at this stage to discuss whether the Commission will have the political will and determination to enforce these measures. Regional marine environmental conventions are already part of Community law. In the past, the Commission could have led a more vigorous enforcement policy with regard to the marine environment, but did not dare to do so. The judgment of the Court of Justice in case C-239/03[73] confirmed the EC's enforcement competence by stating that the provisions of an international convention ratified by the EC form part of EC law.
5. The proposal falls short of the announcement for a comprehensive strategy on the marine environment, mentioned above. Discussions appear to concentrate on the proposed directive, and no measures are announced to transform that strategy into concrete action.

In June 2006, the Commission published its Green Paper on maritime policy.[74] Its objective is "to launch a debate on a future Maritime Policy for the EU that treats the oceans and seas in a holistic way," and seeks "in the wider maritime sphere, to stimulate growth and jobs under the Lisbon agenda in a sustainable manner that ensures the protection of the marine environment." The Thematic Strategy on the marine environment is briefly mentioned, but not discussed. Thus for the purposes of this contribution, the Green Paper is not relevant.

V. Conclusions

The EU does not presently have an integrated, overall policy for the protection of the marine environment. The Thematic Strategy, which took the form of a framework directive, is not likely to significantly change this situation as it builds upon the existing regional structures for the protection of the European seas. As a consequence, progress can be considered relatively satisfactory in the North Sea and the Baltic Sea (it has not been the intention of this contribution to assess that work). The situations in the Mediterranean Sea and the Black Sea will only significantly improve, if future EC legislation is vigorously enforced. Past experience shows that this may not be very likely.

---

72 Article 211 EC Treaty.

73 Case C-239/03 (Note 5 supra).

74 Commission, Communication: Towards a future maritime policy fort he Union: a European vision fort he oceans and seas, COM (2006) 275 of 7 June, 2006.

Who then shall clean up European oceans, the EC or the regional marine conventions? The EC appears to aim at an answer "the regional conventions; if they do not succeed, the EC." Relying on regional marine conventions to curb pollution may be wise in view of the distribution of power within the EC. However, the success of this strategy depends on the EC becoming active in monitoring the elaboration and application of environmental measures in the Mediterranean and Black Sea. Up to the present, the EC has not had the necessary political determination for such a policy. One can speculate whether the EC will develop this determination. With regard to marine protection, Europe is "poor in action, but strong in reflection."[75]

This paper does not discuss the contribution of the European Union to the global problems regarding oceans, because there is practically no such contribution: the EC acts as a regional player and has not actively participated in the global discussions on the laws and policies of the seas.

[75] Paraphrasing Hölderlin's famous statement "tatenarm und gedankenvoll".

HELCOM's Contribution to the Prevention of Marine Pollution

Anne Christine Brusendorff

I. Introduction

For more than three decades, the Helsinki Commission (HELCOM) has acted as the main environmental policy-maker for the Baltic Sea area by developing specific measures to protect and conserve its unique marine environment. The Commission, working through intergovernmental co-operation between all the coastal countries, has produced many environmental gains over the past 30 years. These gains validate the belief that the deterioration of one of the most polluted seas in the world can be arrested and the state of the marine environment improved.

The work and the achievements of HELCOM can be divided into three periods, each strongly influenced by the geo-political circumstances in the Baltic region, from the Cold War era, through to the first and the second EU enlargements.

II. The Beginning

The evident deterioration in the health of the Baltic Sea has been a growing concern since the late 1960s. An increasing awareness that the Baltic Sea is not well, that it cannot survive on its own, and that national measures alone are insufficient to protect its marine environment led the then seven coastal countries to adopt the Convention on the Protection of the Marine Environment of the Baltic Sea Area (Helsinki Convention), signed in 1974.

It was a historical turning point when all of the Baltic Sea states, despite their political and economical differences, agreed in an unprecedented joint effort to protect their common sea. The Convention was also unique in its comprehensive approach to protection of the marine environment; it addressed all sources of marine pollution, be it from sources based on land, at sea or in the air, and also worked on establishing co-operation to improve responses to accidents at sea.

The comprehensiveness of the Convention, together with the limited area that it covers – demarcated by both geographical and ecological factors – enabled more focus on solving complex environmental problems affecting the common sea. The states surrounding the Baltic Sea were naturally drawn together both by their proximity to each other and also by their shared dependence on the semi-enclosed sea.

Although the Convention established internationally legally binding obligations to be undertaken by each Member State, it primarily created the legal basis for a close and permanent co-operation among the Member States. The Helsinki Commission, the governing body of the Convention, would carry out this work.

The creation of the Convention was a political achievement and provided the basis for later environmental improvements by the organisation. Among these achievements are:

- an increased and deepened scientific knowledge of the state of the Baltic Sea and factors affecting it, obtained through harmonised methods used by all states when collecting data on pollution loads stemming from land-based activities and when monitoring the state of the Baltic Sea marine environment. This makes up the foundation of HELCOM's work, enabling assessments of the need for further actions and, thus, serving as supportive information to decision-makers;
- the application of a "substance-wise" approach whereby HELCOM has made Recommendations limiting or eliminating the use of specific substances recognised as harmful to marine ecosystems; and
- the focus on the prevention of pollution from ships and co-operation in case of accidents at sea. While the former, due to the international nature of shipping, requires the adoption of appropriate international requirements and their effective and harmonised implementation by the Baltic coastal countries, the latter is more of a regional matter. To ensure a prompt response to accidents at sea, it is important to assess the adequacy of regional response and emergency capacities as well as to encourage regional co-operation.

III. Expansion of the "Convention Area"

The second era in the history of HELCOM started in 1992. Political changes in Europe and developments in international environmental and maritime law led to a revised Convention signed by all states bordering the Baltic Sea, and the European Commission.

Fundamental new principles and approaches are characteristic of this second era:

- the expansion of the so-called "Convention area" that made not only the sea itself, but also inland waters, subject to protection;
- the enlargement of the area in which the Baltic coastal countries are committed to implement the provisions of the Convention and pertinent Recommendations (the "Area of Application"). This area was enlarged to include the catchment area of the Baltic Sea in addition to the sea area;
- a recognition of the need to mobilise financial resources and of the need for co-operation and co-ordination between HELCOM and the International Financial

Institutions. This work was expanded to include not only Baltic coastal countries but also other countries in the Baltic Sea catchment area, *i.e.* Belarus, Ukraine and the Czech and Slovak Republics. A new approach was taken that used pre-feasibility studies initiated by the International Financial Institutions to identify "pollution Hot Spots" responsible for a major part of the pollution to the Baltic Sea, and ensured this identification would lead to the design of projects capable of attracting funding;

- an extension of the work of HELCOM to cover also nature conservation and biodiversity issues as well as the sustainable use of the natural resources of the Baltic Sea area;
- a shift towards a sector-wise approach to addressing land-based pollution from point and diffuse sources, and also a change towards promoting Best Available Techniques and Best Environmental Practices rather than setting limit values. In HELCOM's work on land-based pollution sources, emphasis is placed on the harmonisation of measures with those taken in the EU and the sister organisation of HELCOM in the North-East Atlantic, OSPAR.

IV. Encouraging Results

Thanks to joint endeavours in the Baltic region encouraging results have been achieved, indicating the rate of degradation of the Baltic Sea has slowed, and that the environmental situation has somewhat improved. As a direct result of HELCOM efforts, many positive improvements in the state of the Baltic marine environment have been observed over the past few decades. Since the beginning of the 1980s, the Helsinki Commission has been working to improve the Baltic marine environment, largely through some 200 HELCOM Recommendations. More than 40 HELCOM Recommendations concern measures limiting pollution originating from the land; either from point sources such as industrial plants and municipal wastewater treatment plants, or from diffuse sources such as traffic and farmland. By implementing these Recommendations, coastal countries have significantly reduced discharges of organic pollutants and nutrients from point sources. Furthermore, numerous projects related to wastewater treatment, pollution control, waste management, maritime safety, biodiversity conservation, and the banning of toxic substance have been carried out.

Nutrient pollution in the Baltic Sea has remained a serious problem and an important political issue since the late 1980s when HELCOM environment ministers set the 50% reduction target for nutrient inputs. The results of a recent study conducted to assess progress of the strategic goals of the 1988 HELCOM Ministerial Declaration regarding nutrient load reductions show much has been achieved in reducing loads from point sources such as municipal and industrial wastewater treatment plants. Almost all of the coastal countries have managed to reach the target of a 50% reduction in phosphorus loads from point sources. However, the results also indicate that measures taken to reduce nutrients originating from agriculture have fallen short of their 50% reduction target. Further implementation of load reduction measures will support reduction of nutrient loads from agricultural sources. Nevertheless, it has been shown that there is a considerable time lag between the implementation of agricultural water protection measures and their effects in water bodies.

Since the early 1990s, a 20-25% overall reduction in the emissions of oxygen-consuming substances (BOD) has been achieved. Great progress has been reported in the reduction of emissions of organo-halogen compounds such as toxic dioxins and furans. Significant reductions in atmospheric nitrogen deposition and inputs of contaminants such as PCB, DDT, mercury and lead are also being accomplished through strict HELCOM regulations. For example, during the period 1996-2000, annual air emissions of heavy metals in the coastal countries decreased by 26% for cadmium, 15% for mercury and 10% for lead. The same is true for dioxins, although their concentrations in Baltic Sea fish still exceed the EU food safety limits. HELCOM Recommendations have resulted in stricter controls imposed on industry, such as compulsory permits for industrial emissions.

Since 1992, 81 of the 162 major pollution "Hot spots" in the Baltic Sea region have been successfully eliminated. This has substantially contributed towards overall pollution load reductions in the Baltic Sea catchment area. Baltic Sea Protected Areas, which serve to protect and restore sensitive ecosystems as well as threatened fauna and flora, have also been established under HELCOM Recommendations.

HELCOM has developed special legislation to prevent pollution from ships, including measures to eliminate illegal discharges by ships into the Baltic Sea, and to ensure the safety of navigation. The Helsinki Commission has also established joint monitoring of the state of the marine environment and carried out regular assessments of its quality status as a pre-condition for evaluating the need for additional protection measures.

All these achievements have resulted in fewer beaches closed for bathing, reduced numbers of illegal oil discharges, the recovery of seal and white-tailed eagle populations, the restoration of wild salmon populations, etc. Despite remarkable progress in past years, the overall state of the Baltic Sea remains unsatisfactory. Many challenges lie ahead of us in our efforts to restore and protect the marine environment of the Baltic Sea.

V. Entering the New Era

HELCOM entered a completely new era in 2004, when Estonia, Latvia, Lithuania and Poland joined the club of Baltic EU Member States. Russia remained as the only non-EU country among the Contracting Parties to the Helsinki Convention. In this radically changing political environment, a large part of the Baltic Sea and its catchment area is now covered by EU-regulations. Such regulations include the European Water Framework Directive and the European Marine Strategy. Eight of HELCOM's nine Contracting Parties have experienced increased workloads due to the fact they must now contribute to and implement EU regulations – even with some decision-making powers delegated to Brussels. As a result, HELCOM's role in the region is now more specific.

In this new role HELCOM will undertake four specific tasks to protect the marine environment:

- first, to improve the ability to identify environmental problems in the Baltic Sea region, and to relate these to quantifiable objectives for a healthy Baltic Sea. These will serve as the foundation for assessing the needs for additional actions and provide supportive information to decision-makers in international fora such as the EU, HELCOM as well as at national levels;
- second, to strengthen indispensable links with Russia and to facilitate co-operation between Baltic EU Member States and Russia through joint HELCOM activities to achieve further environmental protection in the Baltic Sea region;
- third, to maintain HELCOM's important role in strengthening co-operation between HELCOM and the non-HELCOM states in the catchment area, *i.e.* Ukraine and Belarus (also non-EU states) as well as the Czech Republic; and
- finally, to intensify co-operation with other Marine Commissions, aiming for comparable decisions for comparable issues. Examples of this are the joint work programme between HELCOM and OSPAR on marine protected areas and the co-operation between HELCOM, EC, and the Black Sea Commission to address the problem of eutrophication.

VI. Identifying and Addressing Key Environmental Problems

With regard to the first task, the identification of environmental problems and future actions needed for achieving a healthy Baltic Sea, HELCOM has been collecting and disseminating environmental data for over 25 years. This has been accomplished by assessing pollution loads reaching the Baltic Sea and evaluating the state of the Baltic marine environment. What has been lacking in this process is the quantification of the overall goal of setting boundaries for the state of the Baltic Sea. We also need to improve our understanding of the interconnectivity between specific human activities and their effects on the health of the Baltic Sea, and couple this with an assessment of which further measures are needed.

This work is now under development. HELCOM is making progress on its identified environmental priorities of curbing eutrophication, reducing hazardous substances, improving maritime safety, and conserving nature, and biodiversity. Indicators, with target levels are now being established that will allow us to measure the extent to which we are achieving these goals.

VII. Cutting nutrient loads

Eutrophication, the over-enrichment of the Baltic Sea by nutrients, negatively impacts Baltic Sea ecosystems and affects our use and enjoyment of the sea. Examples of direct and indirect effects of eutrophication include intensive plankton blooms, reduced water clarity, and replacement of permanent littoral vegetation with annual vegetation at reduced maximum growth depth, followed by effects on fish, fauna and in the worst cases by oxygen deficiencies in the sea bottom. Consequently, an important aspect of HELCOM's work is to reduce human induced eutrophication to desirable levels. This naturally raises the question: "What are desirable levels?"

In an attempt to define such levels, we are presently determining sub-regional background values for the open sea areas of the Baltic – keeping in mind the need for close co-ordination and co-operation with similar work undertaken for coastal waters under the EU Water Framework Directive. It is not possible to turn back time; we will never have a Baltic Sea of the 1800s. However, background values resulting from natural processes can be used as a reference when aiming for good ecological status. Using these background values will assist in determining target levels that define a minimum desirable status for our sea – a sea with good ecological status, a sea with diverse biological components functioning in balance, and a sea that supports a wide range of sustainable economic and social activities.

Having determined a desired status for our sea, the forces behind the problems should be addressed and, in the case of eutrophication, the levels of nutrient inputs that can be tolerated by the marine ecosystem defined. Nutrient loads must be linked with the activities carried out on land, including the considerable airborne nitrogen inputs originating from activities outside the catchment area such as emissions from distant land-based sources and shipping in the North Sea area. HELCOM can then evaluate priorities for actions by linking catchment input models and airborne nitrogen deposition models to environmental effect models, thereby forecasting and predicting the effects of various reductions in inputs to the status of the sea.

With the knowledge that agriculture is one of the most significant contributors of waterborne nutrients to the Baltic Sea (almost 60% of total nitrogen and 50% of total phosphorus input) HELCOM will elaborate on a thematic assessment report on eutrophication. The objective of this assessment is to evaluate the consequences of various policy scenarios, including alternative agricultural policies of the nine Baltic coastal countries, and to use this information to e.g. provide joint input to the mid-term revision of the EU Common Agricultural Policy in 2009.

VIII. Actions on Hazardous Substances

The gradual pollution of the Baltic marine environment by hazardous substances has seriously threatened the marine ecosystem. Hazardous substances can accumulate in the marine food web to levels toxic to marine organisms. Predators are particularly affected through damage to their endocrine and immune systems, and even human health is at risk.

HELCOM's monitoring activities have indicated that the loads of some hazardous substances have reduced considerably over the past 20-30 years. The last pollution load compilation shows that during the period 1994-2000 riverine loads of heavy metals decreased for almost all coastal countries. However, heavy metal concentrations in the Baltic Sea are still many times higher than in the northern Atlantic, and have not decreased since the 1990s. The best news is that clear decreases in lead concentrations have been observed in herring in most areas, probably due to the significant reduction in the use of leaded fuel.

During the past 50 years, large amounts of other hazardous substances, such as persistent organic pollutants (POPs), have been released into the Balt Baltic Sea from industries, farmlands and waste deposits. Airborne emissions and hazardous substances that have accumulated in sediments over past decades continue to be released into the Baltic Sea, however, direct discharges into waters have dropped significantly.

One of HELCOM's main priorities is to phase out all emissions, discharges and losses of hazardous substances by 2020, so that we can pass on a healthy Baltic Sea to future generations. The goal is to reduce concentrations of man-made hazardous substances to zero and lower concentrations of naturally occurring substances to their natural background concentrations.

From 1999 to 2002, the HELCOM Hazardous Substances Project compiled available data on sources, pathways, markets and the legal situation relating to selected hazardous substances, in order to assess the exposure situation and to identify suitable cost-effective measures to eliminate their use. The project team prepared special documents covering mercury, cadmium, short-chained chlorinated paraffins, nonylphenol and nonylphenolethoxylates, dioxins and PCBs. These guidance documents have been designed to help policy makers choose the most efficient instruments and measures to eliminate emissions, discharges and losses of hazardous substances.

IX. Ensuring Shipping Safety and Response Capacities

Another HELCOM priority is to ensure the safety of shipping, giving special attention to the transportation of oil and building response capacity to accidents at sea. The Baltic Sea is an area of heavy ship traffic; there are approximately 2000 ships at sea at any time, accounting for 15% of the world's cargo transportation. Forecasts indicate that due to economic growth, especially in the eastern part of the region, the amount of cargo shipped on the Baltic Sea will double by 2015 from 500 million tonnes to 1,000 million tonnes annually. In particular, oil transportation is expected to increase, especially in the Gulf of Finland, due to the construction and expansion of Russian oil terminals. Forecasts indicate that by 2010 the total amount of oil transported in the Baltic Sea will increase to 190 million tonnes a year. It is estimated that this increased oil transportation will raise the risk of an oil-spill involving over 10,000 tonnes of oil by 35% for the whole of the Baltic Sea, and by 100% for the Gulf of Finland.

On average, more than 60 ship accidents occur each year. While only a few of these accidents result in oil pollution, even a single incident can devastate the marine environment. To address this threat, the Contracting Parties to the Helsinki Convention co-operate in a variety of ways and HELCOM has adopted various measures to ensure the safety of navigation. These measures include development of relevant legislation, exercises, and operational co-operation during simulation exercises as well as during incidents.

In 2001, HELCOM made efforts to substantially improve the safety of navigation following the biggest oil spill in the Baltic Sea in 20 years. An oil tanker called the Baltic Carrier collided with another vessel spilling 2,700 tonnes of oil near the coast of Denmark, causing widespread pollution of the shore. An Extraordinary HELCOM Ministerial Meeting adopted an extensive package of measures to improve the safety of navigation and to address the adequacy of emergency and response measures in the Baltic Sea area, thereby acknowledging that it will never be possible to totally eliminate the risk for a new accident.

HELCOM recently finalised a very important project to further improve the safety of navigation in the Baltic Sea. On July 1, 2005, HELCOM officially launched an Automatic Identification System (AIS) for monitoring real time maritime traffic in the Baltic Sea. The AIS is an automatic VHF radio-based system that enables identification of the name, position, course, speed, draught and cargo of every ship of more than 300 gross tonnes sailing on the Baltic Sea, and displays all available data over a common background map of the region. With this HELCOM system, it will not only be possible to regularly monitor maritime traffic, but also to elaborate statistics on the nature and extent of shipping and the amount of cargo being transported in the Baltic Sea area. Thus, it provides the basis for future risk assessments and identification of needs for additional measures to prevent collisions and improve navigational safety. In addition to the adoption of the AIS, HELCOM has recently adopted several measures to address operational safety requirements for ships sailing in icy conditions, an area where no other international regulations currently exist.

To ensure a swift national and international response to maritime pollution incidents, several kinds of exercises are regularly conducted under the HELCOM flag. The most famous are the BALEX DELTA exercises, which have been conducted annually since 1990 to test the alarm procedures and response capability of the coastal countries in case of major accidents and international response operations. At present, HELCOM countries have more than 30 sea-going response vessels located around the Baltic Sea. They have necessary equipment, capacity, and trained crew and, in principle, can reach any place in the Baltic Sea within 6-48 hours.

HELCOM countries conduct aerial surveillance flights, including bilateral and multilateral surveillance activities, not only to detect pollution incidents, but also to prevent illegal oil discharges from ships. Statistics show there are presently about 400 illegal oil discharges a year in the Baltic Sea. Despite increased surveillance, there has been a significant reduction in the number of observations since HELCOM initiated the implementation of a comprehensive set of measures to reduce pollution by ship generated waste (known as the Baltic Strategy for Port Reception Facilities for Ship Generated Wastes and Associated Issues) in the late 1990s.

Apart from focussing on the implementation of existing measures to improve maritime safety and uphold and improve the existing response co-operation, in the near future HELCOM will look into:

- the need and possibility of establishing pilotage in high-risk areas;
- the designation and exchange of information on places of refuge, including legal matters related to liability and compensation; and
- ensuring adequate emergency capacity, including emergency towing, lightering and fire fighting capacity, in order to act as early as possible in the case of an accident, thereby reducing possible impacts to the marine environment.

X. Promoting the Conservation of Baltic Sea Biodiversity

Despite the small number of species and relatively simple food chains in marine communities, the Baltic Sea is peculiar and unique in its biodiversity. Its simplicity renders it particularly vulnerable, and perhaps requiring of more care and attention than is normally needed to preserve oceanic biodiversity.

Recognising the need to preserve and conserve Baltic Sea biodiversity, common goals must be set that define the desirable level and state of biodiversity for the Baltic Sea. These goals must define the distribution and abundance of species as well as the general health and structure of marine communities. HELCOM is currently trying to operationalise our common objective of a healthy and viable Baltic Sea – characterised by populations of native species, with no decline in natural biodiversity, restored and preserved natural coastal landscapes and seascapes as well as viable ecosystems in and around the Baltic Sea. In order to assess the success of HELCOM's work, indicators are being developed that will enable us to measure the extent to which we are achieving our common goals. These indicators are being developed using the Ecosystem Approach, which integrates the combined effects of all human activities, i.e., inputs of hazardous substances, human induced eutrophication, and effects of maritime activities. Indicators under development are aimed at detecting and measuring:

- whether we have pollution-related health problems among sea creatures, e.g., reproductive disorders in seals;
- whether all fish caught in the Baltic Sea are safe for human consumption, recalling that e.g. some Baltic fatty fish have higher concentrations of dioxin than the maximum level suggested for human consumption under EU regulations; and
- whether we are combating eutrophication and reaching our objectives of restored water clarity, no unnatural oxygen depletion and no exceptional algal blooms – all factors which influence amongst other things coastal fish stocks and benthic communities.

In 1994, HELCOM initiated the establishment of 62 Baltic Sea Protected Areas (BSPAs) which were set up to protect threatened habitats, fauna and flora. HELCOM is presently assessing the implementation status of these protected areas and determining whether they constitute an ecologically coherent network. Such a network must protect unique areas and special features, and allow biotopes and species to recover from adverse human impacts. This network is being established by compiling information on the type and degree of threat to different marine and coastal biotopes, and by determining the distribution of key, threatened species. The results so far indicate that there is an urgent need to continue implementing conservation measures in the identified BSPA and to take further steps to include 24 proposed offshore BSPAs into the network.

XI. The Road Ahead

Although HELCOM is dedicated to ensuring the obligations, and thereby aim, of the Helsinki Convention are met, there is also an added dimension to our activities. The European Marine Strategy foresees an Action Plan for each eco-region, including the Baltic Sea. HELCOM is in a unique position to deliver on such a plan because of the collaborative relationships already established between the countries in the Baltic Sea catchment area. We are also in the unique position to ensure that the special characteristics of the Baltic Sea are fully accounted for in European policies. Although HELCOM has been working on these issues for more than three decades, the HELCOM Baltic Sea Action Plan will differ on some essential points, *i.e.*:

- all stakeholders, from international organisations and governments, through industries and NGOs, and right down to individual citizens should have an active say in the drawing up of the HELCOM Baltic Sea Action Plan;
- this time we shall define our common vision of a healthy Baltic Sea by using ECOLOGICAL OBJECTIVES. Together we shall agree on ecological objectives for the Baltic marine biodiversity as well as for each of our three main environmental priorities: combating eutrophication, curbing inputs of hazardous substances, and ensuring maritime safety. These ecological objectives will be measurable by indicators with specific targets; and
- the Plan will be implemented via targeted and cost-effective measures.

We will be able to make our environmental policy not only a promoter of a healthy environment but also a driving force for growth and employment. However, the choices we make have to be the choices of our society. It is essential that we develop a common vision of the future shared by the whole range of stakeholders: from the older to the younger generations and from those representing the private and the public sectors.

Further successes, such as those observed during the last three decades, will largely depend on how all the coastal countries and involved stakeholders co-operate to achieve a healthy Baltic Sea environment. The overall state of the Baltic Sea can only be further improved through our combined efforts and integrated actions. HELCOM has a very important role in ensuring the effectiveness of these combined efforts.

Transport of Hazardous and Noxious Goods by Sea – The IMDG Code

Meltem Deniz Güner

I. Introduction

It is estimated that more than 50% of packaged goods and bulk cargoes currently transported by sea can be regarded as dangerous, hazardous or harmful to the environment.[1] These cargoes include products transported in bulk, such as solid or liquid chemicals and other materials, gases, and products needed and produced by oil refineries. Some of these substances, materials and articles are dangerous or hazardous from a safety point of view, and are also harmful to the marine environment. Others are only hazardous when carried in bulk and some may be considered as harmful to the marine environment. Between 10-15% of cargo transported in packaged form falls within this category.

The risks posed by the increased carriage of hazardous, dangerous, and noxious goods by sea have resulted in progressive formulation and adoption of international technical standards to promote maritime safety.[2] Moreover, the public's growing environmental awareness and concern regarding the costs of maritime casualties both at sea and in port have increasingly given rise to consideration of issues of liability and compensation regarding damage caused dangerous or hazardous goods.

Hazardous goods are concern at the risk maritime movement of such goods creates.[3] Government is concerned as rules governing safety carrying with them penalties for breach. Industry is concerned hence special rules contained in the carriage convention. Concerns also exist for marine environment and for those suffering from transportation even if entirely unconnected with it. Therefore, legal

---

1 The Safe Transport of Dangerous, Hazardous and Harmful Cargoes by Sea: (1990) 25 Europ.Transp. L. 747.

2 *De Bievre*, Liability and Compensation for Damage in Connection with the Carriage of Hazardous and Noxious Substances by Sea: (1986) 17 J. Mar. L. & C. 61.

3 *Jackson*, Dangerous Cargo-Legal Overview, Public Regulations & Private Liability, Papers of One Day Seminar, Southampton University (1981) A1.

framework within which industry work covers the area of public and private law, of State concern for safety and environment, of individual concern for liability.

The terms hazardous, dangerous, harmful, or noxious are used to describe such goods. Generally, *dangerous* is used in relation to safety, *hazardous* is used regarding health, and *harmful* and *noxious* are used with respect to environmental effects. However, these distinctions are not always clearly made. Sometimes *hazardous* is used in relation to safety or to describe potential environmental effects. The terms, particularly hazardous and dangerous, have been used comparably in international agreements and national legislation. A clear differentiation is related to the different objectives and definition in the different legislative context.

II. Maritime Safety

Transport of dangerous goods by sea is regulated to prevent human injury or damage to ships and their cargo. The International Convention for the Safety of Life at Sea (SOLAS), 1974 deals various aspects of maritime safety and contains in part A of chapter VII mandatory provisions governing the transport of dangerous goods in packaged form or in solid form in bulk. Regulation VII/1.3 prohibits transport of dangerous goods except in accordance with the provisions of chapter VII part A, which are amplified by the International Maritime Dangerous Goods (IMDG) Code.

III. Marine Environment

Transport of marine pollutants is primarily regulated to prevent harm to the marine environment. The International Convention for the Prevention of Pollution from Ships (MARPOL) 73/78 deals with various aspects of prevention of marine pollution, and contains in its Annex III mandatory provisions for the prevention of pollution by harmful substances carried by sea in packaged form. Regulation 1(2) prohibits the carriage of harmful substances in ships except in accordance with the provisions of Annex III, which are also amplified by the IMDG Code. Although the main concern regarding dangerous goods and IMDG Code is safety, many of the substances listed are also pollutants. Therefore, the Code was extended to deal with marine pollution.

IV. International Maritime Dangerous Goods (IMDG) Code

1. Development of the IMDG Code

Resolution 56, adopted at the 1960 SOLAS Conference, recommended governments adopt a uniform international code for the carriage of dangerous goods by sea. This code would supplement the SOLAS regulations and cover such matters as packing, container traffic, and stowage, with particular reference to the segregation of incompatible substances. It further recommended that the IMO, in-cooperation with the United Nations Committee of Experts on the Transport of Dangerous Goods, continue its studies on such an international code, especially respecting classification, description, labelling, a list of dangerous goods and shipping documents. To carry out this mandate, in January 1961, IMO's Maritime Safety Committee (MSC) established a Working Group on the Carriage of Dangerous Goods (CDG) and group set about preparing the "unified International Maritime Dangerous Goods Code" as envisaged by the 1960 SOLAS.

By November 1965, good progress had been made in composing the IMDG Code, and it was adopted that same year. The IMDG Code is based on the United Nations Recommendation on Transport of Dangerous Goods (Model Regulations), first published in 1957. The Model Regulations serve as a basis for the development of harmonized regulations for all modes of transport in order to facilitate trade and the safe, efficient transport of dangerous materials.

2. Status of the IMDG Code

The IMDG Code was recommended to Governments for adoption or use as the basis for national regulations in pursuance of their obligation under SOLAS and MARPOL Conventions. However, the Code is no longer recommendatory. IMO Resolutions recommend the implementation of technical rules and standards not included in its treaties. These resolutions are non-binding, but it is expected that national regulations conform to these standards, obviously bearing in mind the need to adapt them to the particular circumstances of each case. Technical codes and guidelines included in the resolutions are frequently made mandatory by incorporation into legislation, and in several cases, codes and guidelines initially contained in non-mandatory IMO resolutions are incorporated at a later stage into IMO treaties. The IMDG Code is an example of this process. In order to facilitate multimodal transport of dangerous goods the IMDG Code was incorporated into Chapter VII of SOLAS in 2002, and became mandatory as of January 2004. However, some parts remained recommendatory.[4]

3. Content of the IMDG Code

The IMDG Code provides guidelines for classifying, packaging, marking, labelling, placarding, segregating, and stowing dangerous goods carried in packaged form. Although designed primarily for mariners, the provisions of the IMDG Code affect a number of industries as well as storage, handling and transport services

---

4 Chapter 1.3 (training), section 2.1.0 of chapter 2.1 (class 1- explosives introductory note), section 2.3.3 of chapter 2.3 (determination of flash point), columns (15) and (17) of Dangerous Goods List in chapter 3.2, chapter 3.5 (transport schedules for class 7- radioactive materials), section 5.4.5 of chapter 5.4 (Multimodal Dangerous Goods Form), insofar as the layout of the form is concerned, chapter 7.3 (special provisions in the event of an incident and fire precautions involving dangerous goods, and appendix B.

from manufacturers to consumers.[5] The Code provides guidance to chemical and packaging manufacturers, packers, shippers, forwarders, carriers and terminal operators. Port authorities, and terminal and warehouse companies also consult the IMDG Code to segregate and separate dangerous cargoes in loading, discharge, and storage areas.

The IMDG Code classifies dangerous goods in nine categories according to hazard (or the good's predominant hazard), and has issued an extensive list of dangerous goods.[6] When multiple hazards exist, dangerous goods are classified according to their predominant hazard. The list provided in the IMDG Code is not exhaustive. Dangerous goods that meet the criteria of a single class or division are assigned to that class or division, and the packing group is determined. When an article or substance is specifically listed by name in the Dangerous Goods List (DGL), its class or division, subsidiary risks, and packing group are taken from the list.[7] In cases where dangerous goods meet the criteria of more than one hazard class and are not listed by name in the DGL, the goods are assigned to a class and subsidiary risk on the basis of the precedence of hazard provisions provided in art. 2.0.3 of the Code.[8] The shipper or consignor bears the onus of determining the class of the goods according to provided criteria.

V. Other Instruments Concerning Transportation of Hazardous and Noxious Goods

1. International Code for the Construction and Equipment of Ships Carrying Dangerous Chemicals in Bulk (IBC Code)

Carriage of chemicals in bulk is covered by the regulations in SOLAS 74 Chapter VII and MARPOL 73/78 Annex II. Both conventions require chemical tankers built after 1 July 1986[9] to comply with the IBC Code. The IBC Code provides international standards for the safe transport by sea of liquid dangerous chemicals in bulk, by prescribing the design and construction standards for ships involved in such transport, and specifying the equipment they should carry in order to

---

5 The Safe Transport of Dangerous, Hazardous and Harmful Cargoes by Sea (Note 1 supra) 751.

6 Class 1-Explosives; Class 2-Gases; Class 3- Flammable liquids; Class 4- Flammable solids, substances liable to spontaneous combustion, substances which in contact with water emit flammable gases; Class 5- Oxidizing substances and organic peroxides; Class 6- Toxic and infections substances; Class 7- Radioactive materials; Class 8-Corrosive substances; Class 9-Miscellaneous dangerous substances and articles; and Marine pollutants.

7 IMDG Code 2.0.1.5.

8 IMDG Code 2.0.1.6.

9 Chemical tankers constructed before 1 July 1986 should comply with the requirements of the Code for Construction and Equipment of Ships Carrying Dangerous Chemicals in Bulk (BCH Code).

minimize risks to the ship, its crew and to the environment, having regard to the nature of the products carried.

The basic philosophy is one of ship types related to the hazards of the products covered by the Code. Each product may have one or more hazardous properties including flammability, toxicity, corrosivity and reactivity. The IBC Code lists chemicals and their hazards and gives both the ship and type required to carry that product as well as their environmental hazard rating.

MARPOL Annex II groups "noxious liquid substances carried in bulk" into four categories graded A to D according to the hazard they present to marine resources, human health, or amenities.[10] IBC Code has been incorporated into SOLAS and MARPOL Conventions.

2. International Code for the Construction and Equipment of Ships Carrying Liquefied Gases in Bulk (IGC Code)

The IGC Code applies to gas carriers constructed on or after 1 July 1986.[11] Like the IBC Code, the purpose of the IGC Code is to provide an international standard for the safe transport by sea in bulk of liquefied gases and certain other substances. The Code prescribes the design and construction standards of ships involved in such transport, and the equipment they should carry, so as to minimize the risk to the ship, its crew and to the environment, having regard to the nature of the products involved.

Again the basic philosophy is one of ship types related to the hazards of the products covered by these codes, each of which may have one or more hazard properties. Severe collisions and strandings could lead to cargo tank damage and uncontrolled release of the product.[12] Such releases could result in evaporation and dispersion of the product and, in some cases, could cause brittle fracture of the ship's hull. The requirements in the Code are intended to minimize these risks to the extent practical, utilizing present knowledge and technology.[13] The IGC Code is mandatory under SOLAS 74.

3. The Code of Safe Practice for Solid Bulk Cargoes (BC Code)

The BC Code was adopted in 1965. Its primary aim is to promote safe stowage and shipment by:

---

[10] Revised Annex II "Regulations for the Control of Pollution by Noxious Substances in Bulk" was adopted in 2004. It includes a new four-category – x, y, z, and other substances- categorization system for noxious and liquid substances. The revised Annex is expected to come into force on 1 January 2007.

[11] Gas carriers constructed before that date should comply with the requirements of the Code for the Construction and Equipment of Ships Carrying Liquefied Gases in Bulk or the Code for existing Ships Carrying Liquefied Gases in Bulk.

[12] A further possible hazard may arise owing to the products being transported under refrigerated or pressure conditions.

[13] <www.imo.org/Environment/mainframe.asp?topic_id=995>.

- highlighting the dangers associated with the shipment of certain types of bulk cargoes,
- giving guidance on the procedures to be adopted when the shipment of bulk cargoes is contemplated,
- listing typical materials current shipped in bulk, together with advice on their properties, and
- describing test procedures to be employed to determine various characteristics of the materials to be carried.

The BC Code is still recommendatory, but is expected to become mandatory by 2011.

VI. Transport

The parties to a contract for the transport of goods by sea, whether in the form of a bill of lading or in one of the charterparty forms, customarily negotiate their terms with commercial and legal considerations in mind.[14] Due to the paramount importance of parties' liability for risks to which vessels and their cargo might be exposed during the course of the contract, shippers do not have unlimited freedom as to what may be transported by sea.[15] Therefore, the description of the cargo for shipment is a crucial component of the negotiations between the parties to any shipping contract. Many of the charterparties and bills of lading contain dangerous cargo exclusion clauses.[16] Dangerous goods have been the subject-matter of disputes over hundred years.

1. What is Meant by "Dangerous"?

At common law, and under Article IV (6) of the Hague/Hague-Visby Rules, the shipper may be liable to the carrier in respect of losses it has sustained by loading cargo with dangerous characteristics. Art. IV (6) of the Hague/Hague-Visby Rules provides that "Goods of an inflammable, explosive, or dangerous nature ..." and it is obvious that goods listed in the IMDG Code and other codes within the limit of this expression.[17]

---

[14] *Girvin*, Shipper's Liability for the Carriage of Dangerous Cargo by Sea: (1996) LMCLQ 487, 489.

[15] Ibid.

[16] Cargo exclusion clauses are more common in time charters than in voyage charters. The reason is that in voyage charters, the cargo to be carried is usually mutually agreed upon. See NYPE 93.

[17] Although carriage of dangerous goods is regulated mainly to prevent injury to persons and damage to ship, there should be no reason not to apply it to trade. Incompliance with the IMDG Code and other codes amount of fault of concerned party. In some cases the IMDG Code and other codes are given binding effect by being incorporated into contracts.

However, goods that are not normally dangerous may become so if their condition on loading is unsound. It is difficult to draw a distinction between goods of a dangerous character and goods that merely suffer from an inherent vice , however, this difference remains important.[18] The Hague/Hague-Visby Rules Art. IV (2) m provides that "Neither the carrier nor the ship shall be responsible for loss or damage arising or resulting from ... inherent defect, quality, or vice of the goods". This provision constitutes a defense to a claim for damages arising out of loss or deterioration in the course of carriage, however it does not itself involve any breach of duty by the shipper, or confer any right of recourse on the carrier. Therefore, it is asserted that, in principle, where the condition of the goods on a shipment is such that they are liable to cause injury to persons, damage the ship or other goods, or seriously delay the voyage, such goods fall within the category of dangerous goods.[19] Accordingly, the term *dangerous* is given a broad meaning and is not restricted to "inflammable or explosive".[20] Consequently, a cargo of rice[21] or groundnut[22] may fall within the meaning of *dangerous* under this provision. How-

18 *Cooke/Young/Taylor*, Voyage Charters (2001) 152.

19 Ibid.

20 In *Chandris v. Isbrantsen-Moller turpentine*, highly inflammable solvent was carried despite the fact an express clause in the charterparty existed which prohibited the shipment of "acids, explosives, arms, ammunition or other dangerous cargo". The words "other dangerous cargo" was refused to qualify by reference to the *ejusdem generis* rule. *Chandris v. Isbrandtsen-Moller* [1951] 1 KB 240. The House of Lords took the similar approach in the *Ginannis NK*. It was held that the word "dangerous" in Article IV(6) of the Hague Rules was not qualified by reference back to the preceding words, namely "explosive or inflammable". [1996] 1 Lloyd's Rep. 337 at 341.

21 In *Mitchell Cotts v. Steel* [1916] 2 KB 610, a ship was chartered on a voyage from Basra to Alexandria with a cargo of rice. After the voyage had commenced, the charterers asked the owners to agree to a change of destination (Piraeus) and owners agreed. The charterers, but not the owners, were aware that rice could not be discharged at Piraeus without permission of the British Government. The ship was detained at Piraeus for 22 days while attempts were made to obtain permission, which was eventually refused. The shipowners claimed damages for detention. The court held that whatever full extent of the shipper's obligations may be, at the very least, shippers must undertake that they will not ship goods likely to involve unusual dangers or delay the ship without communicating to the owner facts which are within their knowledge which indicate there is such risk, if the owner does not and could not reasonably know those facts.

22 A cargo of groundnut pellets was loaded on the Giannis NK. Also loaded in other holds was a cargo of wheat pellets. The ship was travelling to ports in the Caribbean where there were strict quarantine and phytosanitary regulations. At the second discharge port, where the groundnut pellets were to land, the cargo was found to be infested with live khapra beetles. Despite fumigation, this infestation was inherent in the cargo upon shipment, and occurred without the knowledge of the shipowners, the charterers, or the shippers. Because the khapra beetle was voracious consumer of foodstuffs, and thus undesirable, the vessel was ordered by local authorities to dump entire remaining cargo at sea or return it to the country of loading. The remaining wheat pellet cargo was not itself at risk of the infestation

ever, it is questionable whether *dangerous* can be so broadly construed in the light of existing regulations concerning dangerous goods and the intentions of framers of Hague-Rules.[23]

2. Nature of the Shipper's Liability

Common law imposes an obligation on a shipper of cargo not to ship dangerous cargo without giving notice to the carrier.[24] The reason for imposing such an obligation is clearly to give carriers the opportunity to refuse to carry the goods, or to take the necessary precautions to protect their ship and the cargo of other shippers on board.[25] The shipper discharges this obligation to carrier when notice is given.[26] If the carrier is aware of the dangerous character of goods, the shipper will be under no obligation to give notice.[27] The shipper's obligation is *absolute*; it is not dependant on whether the shipper himself knew or should have known the dangerous character of the cargo.[28]

Art. IV.6 of the Hague/Hague-Visby Rules provides that a "shipper of dangerous goods shall be liable for all damages and expenses". The rules also state that the shipper's liability is, in general, fault based.[29] It has been questioned whether a shipper's liability arising out of dangerous goods was strict or fault-based. This issue now seems to be resolved. The House of Lords confirmed absolute nature of shipper's liability in *The Giannis NK* holding that Hague Rules Art. IV.6 was a free-standing provision dealing with a specific subject matter.[30] The court also stated that Art. IV.6 was neither expressly nor implicitly subject to

spreading to it, but it was likely that the consequences of the infestation of the ground nut cargo was that it would also have to be destroyed. Indeed, both cargoes were ultimately dumped at sea. The vessel was then extensively fumigated so that she was fit for further trading. The court held that the cargo was dangerous within the meaning of Article IV.6, as it was liable to give rise to the loss of other cargo by dumping it at sea, and to the quarantining of the vessel until after she had been fumigated. What made the cargo dangerous was the fact that the shipment and the voyage were to countries where the imposition of quarantine, and an order for the destruction of the entire cargo were to be expected of as at least a natural and not an unlikely consequence of the presence of the infested goods. *Effort Shipping v. Linden Management (The Giannis NK)* 1 Lloyd's Rep. 337.

23 It seems that intention of this provision to cover chemicals or chemical reactions not any good or any danger; *Sturley*, Legislative History of the Carriage of Goods by Sea Act, Vol. 1 (1990) 272-273.

24 *Brass v. Maitland* (1856) 6 E&B 470, 482.

25 *Girvin* (Note 14 supra) 491.

26 Ibid.

27 Ibid.

28 *Brass v. Maitland* (Note 24 supra) at 482; *Great Northern Railway v. L.E.P Transport* [1922] 2 KB 742; *The Athanasia Comminos* [1990] 1 Lloyd's Rep. 277.

29 Hague/Hague-Visby Rules Art. IV.3 "The shipper shall not be responsible for loss or damage sustained by the carrier or the ship arising or resulting from any cause without act, fault or neglect of the shipper, his agents or his servants."

30 Note 22 supra.

Art. IV.3[31] It imposed strict liability on shippers in relation to the shipment of dangerous goods irrespective of fault or neglect on their part.[32]

Previously, American law was generally interpreted to mean that cargo owners did not warrant that the cargo contained no inherent dangers.[33] The cargo owner's duty was based on either its actual or constructive knowledge of danger.[34] However, in *Senator Linie*[35] the Second Circuit held that where neither the shipper nor the carrier has actual or constructive knowledge of the inherently dangerous nature of shipped goods, the specific language of COGSA §1304(6) makes the shipper strictly liable for damages and expenses arising from their shipment. The Court's decision provides a highly detailed review of COGSA. The Court also considered the interplay between Sections 3 and 6 and stated that: "The specific language and subject matter of COGSA §1304(6) indicate that the latter provision is an exception to, rather than a special application of, the fault-based standard of §1304(3)."[36] The Second Circuit's decision plainly indicates that these two sections impose distinct types of liability; and it appears that a cargo owner could rely on either fault-based liability or strict liability when attempting to make the shipper responsible for damages caused by dangerous cargo.[37] In reaching its decision, the court relied on, and noted its agreement with, the ruling of the House of Lords in the *Giannis NK* and was mindful of the importance of maintaining international uniformity in laws governing the maritime trade.[38]

No dispute arises regarding basis of shipper's liability under the German Law. The German Commercial Code §564b expressly states the shipper is strictly liable when such cargo shipped without knowledge of the carrier or master.

3. Dangerous Goods and New Developments in the Carriage of Goods by Sea

At the request of UNCITRAL, CMI has been over the past few years preparing a draft for a new convention on the contract of carriage by sea. It is UNCITRAL's intention to create, on the basis of this draft, a successor to the Hague/Hague-

---

[31] Ibid. at 342.

[32] Ibid.

[33] For more information see *Wilford/Coghlin/Kimball*, Time Charters (5th ed. 2003) 187; *Bulow*, Dangerous Cargoes – The Responsibilities and Liabilities of the Various Parties: (1989) LMCLQ 342.

[34] *International Mercantile Marine Co. V. Fels,* 170 F.275, 277 (2nd Cir. 1909); *William J. Quilllan,* 180 F.681, 682-684 (2nd Cir. 1910); *Akt. Fido v. Lloyd Brasileiro,* 267 F.733 (S.D.N.Y. 1919); *Sucrest Corp. v. Jennifer,* 455 F.Supp. 371. However, it has also been held that there is an implied warranty given by the shipper of cargo that the shipment is reasonably fit and safe for carriage; *Pierce v. Winsor,* F.Cass, Nos. 11.150 and 11.151 (D.C.C Mass. and C.C.D Mass. 1861).

[35] *Senator Linie GmbH & Co. v. Sunway Line, Inc.*, 2002 AMC 1217, 1231.

[36] Ibid. at 1230.

[37] *Wilford/Coghlin/Kimball* (Note 33 supra) 190.

[38] *Senator Linie* (Note 35 supra) 1246.

Visby Rules and the Hamburg Rules. The Draft Instrument started life within the CMI in 1998.

In the Draft Instrument[39] the distinction between dangerous goods and ordinary goods had been lifted. This distinction was considered out of date, as the notion *danger* has received more relative character today.[40] Inherently dangerous cargo, which is packed according to applicable regulations, and is handled and carried by the carrier in the manner as appropriate for that kind of cargo, is in absolute sense not necessarily more dangerous than ordinary cargo. Under the draft rules the shipper has two main obligations. The first is to deliver the goods to the carrier as agreed upon and ready for the envisaged carriage and capable to withstand it.[41] The sanction for breach of this obligation by the shipper is fault liability.[42] With regard to liability, distinction was not made between dangerous and regular goods. The shipper's second main obligation is to provide the information, instructions and documents that are reasonably necessary for the carriage, the compliance with all the rules connected and proper transport documentation.[43] A shipper in breach of the second obligation was made strictly liable.[44]

However, in its 13th session the Working Group agreed to insert a specific provision in the draft instrument deal with the issue of dangerous goods. Further the provision should make a shipper of goods strictly liable where insufficient or defective information regarding the nature of the goods was provided.[45] It was also decided to broadly define *dangerous goods* in the draft instrument. In doing this other existing international instruments will be taken into account.[46]

VII. Third Party Liability Arising out of Transport of Hazardous and Noxious Goods

1. The International Convention on Liability and Compensation for Damage in Connection with the Carriage of Hazardous and Noxious Substances by Sea (HNS Convention)

The HNS Convention was adopted under the auspices of the International Maritime Organization in 1996. The HNS Convention has not come into force yet, but

---

39 It is available on the UNCITRAL website at <www.uncitral.org>.

40 *van der Ziel*, The UNCITRAL/CMI Draft for a New Convention Relating to the Contract of Carriage by 2002 Sea: TranspR 2002, 265, 272.

41 *Asariotis*, Main Obligations and Liabilities of the Shipper, TranspR 2004, 284, 288.

42 UNCITRAL/CMI Draft Instrument, Art. 30.

43 Ibid. at. 290.

44 UNCITRAL/CMI Draft Instrument, Art.29.

45 WG III A/CN.9/552 at 33-34.

46 The only possible reference was said to be the definition provided in the HNS Convention, but considerable doubts were expressed regarding appropriateness of introducing such a definition in an international trade law instrument.

is expected to come into force in 2008.[47] At present, shipowners' limitation of liability arising from an incident involving the carriage of HNS by sea is governed by the general rules on limitation under the International Convention on Limitation of Liability for Maritime Claims, 1976 (LLMC 76).[48] However, several concerns provided the impetus for legal change and the subsequent adoption of the HNS Convention, including fears of:

- potentially massive claims for environmental damage caused by spillage of chemicals and other non-oil pollutants carried by ships due to the fact the existing conventions on pollution damage mainly concern pollution damage by oil carried in bulk,
- damage other than pollution, such as fire and explosions, caused by oil, chemicals, liquefied gas, or other substances, and
- the risks posed by packaged dangerous cargoes such as these being washed up on beaches.

The regime established by the HNS Convention is largely modelled on the existing regime dealing with oil pollution from tankers, set up under the International Convention on Civil Liability for Oil Pollution Damage 1992 (the "CLC") and the International Convention on the Establishment of an International Fund for Compensation for Oil Pollution Damage 1992 (the "Fund Convention"). The CLC and the Fund Convention cover pollution damage caused by spills of persistent oil from tankers.[49] The main objective of the HNS Convention is to

---

47 One of the reasons for failure to ratify the Convention is that the convention includes agreements between signatory states concerning jurisdiction and the mutual enforcement of judgements. The problem is that EU member states have ceded to the EU so that individual EU states were not able to individually ratify the convention. However, the EU now authorises member states to ratify the convention and requires member states to ratify before June 2006. 2002/971/EC Council Decision of 18 November 2002 authorising the Member States to ratify or to accede the International Convention on Liability and Compensation for Damage in Connection with the Carriage of Hazardous and Noxious Substances by Sea, 1996.

48 The 1996 protocol, which came into force in 2004, substantially raised the amount of compensation payable in the event of incident, and introduced "tacit acceptance" method for updating these amounts.

49 Although the system is similar to the CLC, there are significant practical differences. Regarding CLC, the oil receiver contributors are likely to be relatively few but large, often multinational companies such as the oil majors who generally accept their responsibility for pollution caused by their cargoes and who appreciate the affects of pollution claims on their corporate image. They are also in a position to monitor the ships that carry their cargoes by vetting and inspections. By contrast, contributors to the HNS Fund, and the general Fund particularly, are likely to include medium or small size companies who have not previously felt any responsibility for the carriage of their cargoes, have no influence over the quality of the ships that carry them, or feel that damage caused by their cargoes will not have significant impact on their corporate image. Moreover, The HNS regime is governed by solely one convention.

provide adequate, prompt, and effective compensation for loss or damage arising in connection with the carriage of HNS on sea-going ships.

2. Substances Included in the Definition of Hazardous and Noxious

The HNS Convention defines the concept of hazardous and noxious substances (HNS) by reference to existing regulations and codes.[50] HNS are very varied and include both bulk and packaged cargo.[51] Bulk cargoes can be solids, liquids (including oils), or liquefied gases. The number of substances falling within the category of HNS is very large. Bulk solids are included if they are covered by Appendix B of the BC Code, i.e., if they posses chemical hazards and if they are subject to the provisions of the IMDG Code when carried in packaged form.[52] The effect of classifying HNS in this way is that many of the major bulk solids are excluded, since they either do not pose chemical hazards,[53] or they are classified as materials hazardous only in bulk (MHB).[54] Bulk liquids are considered HNS if they present safety, pollution or explosion hazards; they include organic and inorganic chemicals, vegetable and animal oil, and fat, as well as persistent and non-persistent petroleum oils.[55] As these codes and regulations are amended, the HNS Convention will also be tacitly amended.

---

50 HNS Convention, Art. 1.5.

51 In total HNS cover some 35 000 substances. States may choose not to apply this convention to ships which carry hazardous and noxious substances only in packaged form. Art.5/1b.

52 Draft IMO Guide for Interested Parties on the Workings of the Hazardous and Noxious Substances Convention 1996 (HNS Convention), IMO Legal Committee, LEG/INF.3 at 8.

53 Examples of such non-hazardous chemicals include iron ore, grain, bauxite and alumia, phosphate rock, cement, and some fertilisers.

54 Examples of materials hazardous only in bulk (MHB) include coal, reduced iron or woodchip. The issue of whether or not coal and low-hazard materials carried in bulk should be covered by the HNS Convention was much debated. A clear majority of delegations, especially countries with an extensive import or export of coal, supported its exclusion, because, in their opinion, reliable statistics showed that coal could not cause any damage to the environment or outside the ship. The emission of the methane was not deemed a safety hazard justifying the inclusion of coal, which would substantially increase transport and insurance costs, thus causing serious disadvantages to the national economies of several countries. *Wetterstein*, Carriage of Hazardous Cargoes by Sea- the HNS Convention": (1997) 26 Ga.J.Int'l & Comp. L. 595, 601.

55 Ibid. Although the Convention only covers non-pollution damage caused by persistent oil.

3. Damages Covered by the HNS Convention

HNS incidents pose a significant risk due to the immense and growing amount of international trade in hazardous and noxious substances.[56] Numerous factors influence the magnitude of the damages resulting from an HNS incident. These factors include the type and character of the substances released, the particular characteristics of the casualty, the quantity of the substances released, the physical and meteorological conditions at the time of the incident, the ecological sensitivity of the body of water and coastline to pollution damage, the technical difficulties arising from the incident, the use of the body of water for navigation, recreation or fishing and the population, and type of property located near the incident.[57]

The HNS Convention covers the following damages:

- loss of life or personal injury on board or outside the ship carrying HNS,
- loss of, or damage to property outside the ship ,
- loss or damage caused by contamination of the environment, and
- the cost of preventive measures[58]

As pollution damage caused by persistent oil is already covered under the existing international regime established by 1992 CLC and Fund Convention, such damage is not covered by HNS Convention. However, non-pollution damage, such as fire or explosions caused by persistent oil, is covered by the HNS Convention. Furthermore, the HNS Convention does not apply to damage caused by radioactive materials,[59] or to claims arising from contracts of carriage.

4. Liability under the HNS Convention

a) First Tier – Shipowner's Liability

The HNS Convention provides two tier system of liability. In the first tier, the registered owner of the ship in question is strictly liable to pay compensation following an accident involving HNS.[60] This means the registered owner is liable

[56] *Pawlow*, Liability for Shipments by Sea of Hazardous and Noxious Substances: (1985) 17 Law & Pol'y Int'l Bus. 455, 458-459.

[57] Ibid.

[58] Art. 1.6: The Convention defines *preventive measures* as any reasonable measures taken by any person after an incident has occurred to prevent or minimize damage. Art. 1.6.7: These include measures such as clean-up or removal of HNS from a wreck, if the HNS present a hazard or pollution risk.

[59] Radioactive materials are governed by The Vienna Convention on Civil Liability for Nuclear Damage, 1963; Paris Convention on Third Part Liability in the Field of Nuclear Energy, 1960; Joint Protocol Relating to the Application of the Vienna Convention and the Paris Convention, 1988.

[60] In order to prevent unnecessary litigation, liability is imposed on the shipowner and does not attach to other persons connected with the operation of the ship, unless the damage resulted from the fault of the other persons. A two-tier system of shipowner/shipper liability was proposed. Supporters of two-tier system believed that the risk associated with the maritime shipment of hazardous materials

even in the absence of fault on his part, and all that must be established is a causal link between the damage and the HNS carried on board.

Liability is limited to the shipowner, and excludes his crew, employees or agents, charterers,[61] pilots, salvors, or those taking preventive measures. A shipowner is not liable if he can show: the damage was caused by war, etc.; the damage resulted from a natural phenomenon of an exceptional, inevitable, and irresistible character; third party intent to cause damage; failure of lights or navigation aids; negligence or a wrongful act by the responsible government or authority responsible for maintaining lights or other navigational aids; failure of a shipper to warn the shipowner of the nature of the HNS; and the damage was caused by the person suffering damage through negligence or intent.[62]

Shipowners are normally entitled to limit their liability under the HNS Convention to an amount calculated on the basis of the units of gross tonnage of the ship. The aggregate amount of a shipowner's liability shall not exceed 100 million special drawing rights (SDR).

Strict liability of shipowner, together with the complementary HNS Fund, strengthens the position of claimants.

b) Compulsory Insurance

The owner of a ship that carries HNS is required to take out insurance or maintain other acceptable financial security to cover his liability under the HNS Convention. The HNS Convention requires shipowners' to provide evidence of insurance cover upon the ship's entry into port of any State which is party to the Convention by production of a certificate, regardless of whether the State of the ship's registry is party the convention.

Claimants have a direct right of action against the insurers. Nevertheless, insurers have the right to limit their liability according to the ship's gross tonnage, even though shipowners themselves do not have this right. Insurers can also invoke defenses normally available to the shipowner, with the exception of defences such as bankruptcy, etc.[63]

---

should be shared, because shipping is a joint venture. Furthermore, it was argued that the marine insurance market cannot support increased coverage of shipowners, while shippers have access to other insurance markets since they usually have deeper pockets. Supporters of single-tier system believed that their approach would have been similar to implement and would have avoided other controversial issues. It was also argued that the most casualties result from the vessel operator's negligence and there is little a shipper can do to change the risk of carriage. *Pawlow*, "Liability for Shipments by Sea of Hazardous and Noxious Substances", supra note 56, at 470.

61 Include bareboat charter.

62 HNS Convention, Art.7.

63 HNS Convention, Art. 12.8

c) Second Tier – HNS Fund

If the total admissible claims exceed the shipowner's liability, then the HNS Fund comes into operation. The HNS Fund disburses "top up" compensation when the shipowner or the shipowner's insurer cannot fully pay the costs of the loss or damage. In other words, the HNS Fund will pay compensation in excess of the shipowner's liability, but with a limit of 250 SDR.[64] The HNS Fund also pays compensation if the shipowner is exonerated from liability, or if the shipowner liable for the damage is financially incapable of meeting his obligations.

When fully operational, the HNS Fund will have four accounts:[65]

1. Oil
2. Liquefied Natural Gas
3. Liquefied Petroleum Gas
4. General Account
   1. Bulk solids
   2. Other HNS

The rules do not allow for cross-subsidization among accounts, which means that each account will meet the costs of compensation payments arising from damage caused by substances contributing to that account.

Compensation payments made by the HNS Fund will be financed by levies on receivers of HNS. Receivers[66] of HNS might contribute to one or more of the accounts. The levies applying to individual receivers will be calculated according to the quantities of contributing cargo received above certain thresholds.[67]

---

64 This includes the amount paid by the shipowner or his insurer.

65 During the early existence of the HNS Fund, there may not be sufficient contribution basis in the form of the quantities of HNS received in Member States to set up all the separate accounts. Therefore, initially, the HNS Fund may have only two accounts: a separate account for oil, and a general account with four sectors.

66 The *receiver* is an individual or company who physically receives more than a specified quantity of hazardous and noxious substances in the ports and terminals of a State Party. States are also allowed to establish their own definition of *receiver* under national law. HNS Convention, Art. 1.4.

67 For persistent oil 150.000 tonnes, non-persistent 20.000, LPG 20.000. There is no minimum quantity for LNG.

Part III:

Compensation for Marine Pollution

Origins and Compensation of Marine Pollution – A Survey

Peter Ehlers

I. Introduction

Preventing pollution is the best way to protect the marine environment. Because pollution will never be totally eliminated, compensation for pollution damage is an important form of protection. Compensation is necessary for environmental restoration, but it also functions as a deterrent, and thus, has a preventive effect. The polluter-pays-principle, a main concept in environmental law,[1] embodies the concept of compensation.

When speaking about compensation of marine pollution, pollution from ships automatically comes to mind. The international legal regime dealing with liability and compensation is comprised of four international maritime conventions. Whether these conventions offer the most appropriate regulatory solution is debatable. The regime can be further criticised in view of the fact two of the conventions are not yet in force, and it remains unknown when the necessary quorum for ratifications will be met. So there is enough food for some heavy discussion about the compensation regime for shipping. However, doing so I feel a bit queasy. The emphasis placed on shipping regulations reinforces the impression that shipping is the main cause of marine pollution, and that marine environmental problems can be solved if a perfect and complete compensation regime for shipping is found. This impression is totally wrong. This paper will examine different sources of marine pollution and their environmental effects in order to determine the ade-

---

1 This principle is reflected in Article 3 paragraph 4 Helsinki Convention (Convention on the Protection of the Marine Environment of the Baltic Sea Area, 1992, text published in BGBl. 1994 II p. 1397), Article 2 paragraph 2 (b) OSPAR Convention (Convention for the Protection of the Marine Environment of the North-East Atlantic, text published in BGBl. 1994 II p. 1360); its objective is to channel the costs of prevention and reparation of environmental damage to the person who is in the best position to prevent such damage and to internalize the costs of pollution damage, see Liability & compensation regimes related to environmental damage: Review by UNEP-Secretariat, p. 27 et seq., 103, available from <www. unep.org/depi/liability and compensation.asp>.

quacy of the existing compensation regulations covering all types of marine pollution. Only a rough survey will be presented, and further in-depth study is required. The paper will not focus German domestic law,[2] but instead on international approaches.

II. Sources of Marine Pollution

Marine pollution is defined as the introduction by man, directly or indirectly, of substances or energy into the marine environment, which results or is likely to result in such deleterious effects as:

- harm to living resources and marine life,
- hazards to human health,
- hindrance to marine activities, including fishing and other legitimate uses of the sea,
- impairment of quality for use of sea water and
- reduction of amenities.[3]

Polluting substances enter the marine ecosystem via a number of pathways, in particular:

- riverine runoff,
- atmospheric deposition
- direct discharges from land, and
- activities at sea.

Anthropogenic inputs, the substances introduced into the sea by man, inputs include nutrients, heavy metals, and organic substances such as hydrocarbons. The vast majority of anthropogenic inputs are from land-based sources within the catchment area, originating from municipalities, farmland, commercially managed forests, industrial and energy plants, transport, and other human activities. Pollutants from far away sources enter seas from the air. Emissions and discharges from offshore activities, shipping and fish farms, also enter the sea directly. All inputs adversely affect the quality of seawater and marine life.

Though considerable efforts have been made to reduce the input of nutrients in the sea, eutrophication. will remain an issue of major concern in particular in the Baltic Sea, which is highly sensitive due to its natural conditions. Scientific evidence demonstrates that even if the input of nutrients stopped, existing con-

---

2 Concerning domestic German law see *Erbguth*, Entwicklungen der Umwelthaftung: Ansätze eines Haftungsregimes für die Verschmutzung der Meere im deutschen Recht, in: 1. Rostocker Gespräch zum Seerecht – Aktuelle Probleme der Haftung für Schäden aus der Meeresverschmutzung, Schriften des Deutschen Vereins für internationales Seerecht, Reihe A, Heft 84; cf. also *Bussek*, Schutz der Meere vor Verschmutzung, p. 82 et seq.

3 Article 1(4) UNCLOS (United Nations Convention on the Law of the Sea, BGBl. 1994 II p. 1798).

centrations in sediments and the possibility of their remobilization would adversely affect the marine ecosystem for quite a long time. Eutrophication leads to intense algal blooms, which have many adverse impacts, especially on tourism. It also results in increased oxygen consumption, ultimately causing the formation of toxic hydrogen sulphide, which is a great threat to marine life.

Although heavy metals discharges have decreased, their concentrations in coastal areas remain many times higher than in the open ocean. No wonder that the most encouraging results have been achieved with the reduction of lead inputs through the increasing use of unleaded petrol, but the overall picture remains unclear. For example, cadmium concentrations have declined in the waters, but have increased in the tissues of organisms in some areas.

In the past 50 years, large amounts of persistent organic pollutants (POPs) have been released into waters from different sources, including industrial discharges, runoff from farmland, antifouling paints, and onshore dumping sites. The levels of several POPs, such as DDT, PCB and lindane, have significantly decreased, probably as consequence of strict international regulations. Dioxin concentrations have also been reduced, although in some areas like the Baltic Sea, dioxin concentrations in fish still exceed the EU food safety limits. Some recent measurements show high levels of TBT in marine snails which may have biological effects. Even greater concern is given by the fact that due to the lack of monitoring activities and accurate data regarding the concentration levels of many organic contaminants are still unknown, not to mention possible adverse effects. It is quite alarming that every day new, unknown, hazardous substances are possibly entering the sea.

Oil spillages from ships as a result of illegal discharges of oil residues or caused by an accident are another matter of concern. The same is true for oil spills originating from offshore activities. However, it is important to note that land-based sources are by far the largest contributor to oil inputs. These inputs carry oil in a dissolved form, and are therefore not visible. However, they too represent a large threat to marine life.[4]

III. Damage

As all these polluting inputs have adverse effects on the marine environment they may cause damage in a legal sense, making compensation necessary. When defining the term *damage* two main categories have to be considered. First, the pollution may damage health or property of persons or result in economic loss,

[4] For a detailed description of pollution sources and their impacts on the marine environment see *OSPAR Commission*, Quality Status Report for the North-East Atlantic, 2000; *Helsinki Commission*, The Baltic Marine Environment 1999 – 2002, Baltic Sea Environment Proceedings No. 87; cf. *Ehlers*, The Baltic Sea – Threats and Future Priorities, in: Maritime Safety – Current Problems of Use of the Baltic Sea, European Association of Legislation, ed. by Karpen, Vol. 11 (2005) 13 et seq.

what may be called damage *through* the environment When dealing with economic loss, a further distinction must be made between a direct loss occasioned by property damage, and pure economic loss suffered, e.g., by fishermen or the tourist industry that depend on an unpolluted marine environment. Second, the pollution may cause damage *to* the environment resulting in ecosystem degradation;[5] this damage comprises the costs of preventive measures to minimise the damage, for clean-up measures, measures required to restore natural resources or of acquiring an equivalent and – from a dogmatic view – even the diminution of value of damaged resources. The first category is called *traditional damage*[6] and falls into civil law, as the rights and interests of individual persons are affected. The second category is referred to as *environmental damage*[7] and is strongly influenced by administrative or public law.[8] In such a case public authorities have to take action by requiring the polluter to take remedial measures, public authorities may themselves act, or a third party may take action by making a claim for compensation to recover their costs. The question then becomes to what extent international law provides regulations allowing for compensation of both types of damage caused by marine pollution.

IV. Enabling clauses

A number of international conventions exist requiring states to establish relevant compensation legislation. The most prominent is Article 235 UNCLOS, which requires states ensure recourse is available for obtaining prompt and adequate compensation for damage caused by pollution of the marine environment. This article also calls for co-operation from states to develop international law regarding responsibility and liability for the assessment of, and compensation for damage.

A similar requirement is mirrored in two declarations, which although they are only soft law instruments, provide a strong incentive for future development of international environmental law. Principle 22 of the 1972 Stockholm Declaration of the United Nations Conference on the Human Environment[9] and Principle 13 of

---

5 This distinction is made by *de La Fayette*, Compensation for Environmental Damage in Maritime Liability Regimes, in: International Maritime Environmental Law, ed. by Kirchner (2003) 262; European Commission, White Paper on environmental liability, COM (2000) 66 final, p.16 et seq.

6 White paper (Note 5 supra).

7 *UNEP* (Note 1 supra) 27, defines environmental damage as a change that has a measurable adverse impact on the quality of a particular environment or any of its components, including its use and non-use values, and its ability to support and sustain an acceptable quality of life and a viable ecological balance.

8 See Directive 2004/35/EC, Preamble, paragraph 15, OJ 2004 L 143, p. 56; cf. White Paper (Note 5 supra) 28.

9 Cf. *Balkin*, Some Future Developments in Liability and Compensation for Environmental Damage at Sea, in: The Stockholm Declaration and Law of the Marine Environment, ed. by Nordquist/Moore/Mahmoudi (2003) 437.

the 1992 Rio Declaration on Environment and Development[10] urge States to develop national law regarding liability and compensation for victims of pollution and other environmental damage, and to co-operate to develop international law regarding liability and compensation for adverse effects of environmental damage caused by activities within their jurisdiction, or control to areas beyond their jurisdiction.[11]

These declarations are not restricted to the marine sphere, but cover the environment as a whole. This is also true of two global conventions that establish corresponding enabling clauses. In compliance with the Convention on Biological Diversity[12] the Conference of the Parties shall examine the issue of liability and redress, including restoration and compensation, for damage to biological diversity.[13] This matter became one of the items on the agenda during the negotiation of the Cartagena Protocol,[14] which seeks to protect biological diversity from potential risks posed by living modified organisms resulting from modern biotechnology. However, as no consensus could be reached, only an additional enabling clause was included in the Cartagena Protocol.[15] The Basel Convention[16] also requires the Contracting Parties to adopt a protocol setting out appropriate rules in the field of liability and compensation for damage resulting from the transboundary movement and disposal of hazardous wastes and other wastes.[17] The London Dumping Convention[18] can also be mentioned in this context, as it urges Contracting Parties to develop procedures for assessing liability concerning dumping.[19]

In accordance with UNCLOS and the Stockholm and Rio Declarations numerous regional marine environment protection conventions call upon Contracting Parties to co-operate in developing regulations relating to liability and compensation for pollution damage.[20] These agreements only create somewhat stringent obligations for the Contracting Parties to develop provisions, and are complemented by others that establish specific compensation regulations. The most

10 This principle reaffirms Principle 22 of the Stockholm Declaration of the United Nations Conference on the Human Environment, 1972.

11 The Declarations are available from <www.unep.org/documents>.

12 BGBl. 1993 II p. 1741.

13 Article 14.

14 Cartagena Protocol on Biosafety, BGBl. 2003 II p. 1506.

15 Article 27; an open-ended ad hoc group on liability and redress shall complete its work in 2007.

16 Convention on the Control of Transboundary Movements of Hazardous Wastes and their Disposal, BGBl. 1994 II p. 2703; 2002 II p. 89; 2003 II p. 1626.

17 Article 12; see infra part 8.

18 Convention on the Prevention of Marine Pollution by Dumping of Wastes and other Matter, 1972, BGBl. 1977 II p. 180.

19 Article X; *de La Fayette*, (Note 5 supra) 232, points out that the Contracting Parties have decided that no liability regime is necessary as all dumping of hazardous substances has theoretically been phased out.

20 For details see *UNEP* (Note 1 supra) 94 et seq.

important agreements referring to compensation of marine pollution will be elaborated on below.[21]

V. IMO Conventions

An international legal regime respecting pollution damage caused by ships has been elaborated under the auspices of the IMO. It consists of four international conventions dealing with pollution from oil and hazardous and noxious substances including:

- the International Convention on Civil Liability for Oil Pollution Damage, 1992,[22]
- the International Convention on the Establishment of an International Fund for Compensation for Oil Pollution Damage, 1992,[23]
- the International Convention on Liability and Compensation for Damage in Connection with the Carriage of Hazardous and Noxious Substances by Sea, 1996,[24]
- the International Convention on Civil Liability for Bunker Oil Pollution Damage, 2001.[25]

Among the above, only the first two conventions are in force. The conventions follow a coherent concept. As concerns any details I refer to the presentations made by others later and will confine myself to some leading principles. The term *damage* in this context covers the loss or damage caused outside the ship by contamination and the costs of preventive measures to minimise further damage. Impairment of the environment, other than loss of profit from such impairment, is limited to costs of reasonable measures of reinstatement. The ship owner is strictly liable up to a maximum limit and is required to have insurance. Once this limit is reached, additional compensation for pollution by hazardous and noxious substances or by oil from a tanker is paid from a fund.[26] Additionally, a convention on wreck removal providing compensation for the costs of removing wrecks that are hazardous for the environment is under development.

---

[21] Altogether *UNEP* (Note 1 supra) 18 et seq., lists 54 international agreements relevant for liability and compensation related to environmental damage.

[22] CLC; BGBl. 1996 II p. 670; 2002 II p. 943.

[23] Fund Convention, BGBl. 1996 II p. 685; 2002 II p. 943.

[24] HNS Convention, OJ 2002 L 337, p. 55.

[25] Bunkers Convention, OJ L 2002 256, p. 7.

[26] For more details see *UNEP* (Note 1 supra) 34 et seq., 38, 41; *de La Fayette* (Note 5 supra) 236 et seq.; *Balkin* (Note 9 supra) 438 et seq.

VI. Offshore pollution liability

No international convention exists dealing with pollution damage caused by offshore activities. In the 1970s the North Sea States elaborated a Convention on Civil Liability for Oil Pollution Damage Resulting from Exploration for and Exploitation of Seabed Mineral Resources. The concept was that the operator is strictly liable for damage originating from the installation up to certain financial limits and has to maintain insurance. However, these endeavours ultimately failed, as only three States signed the agreed text and the Convention never entered into force.[27]

Despite the fact this agreement failed, discussions about a convention provoked the oil companies to introduce the Offshore Pollution Liability Agreement (OPOL). OPOL is a private agreement whereby owners and operators of offshore facilities have agreed to a strict liability scheme for the payment of pollution damage claims and the cost of remedial measures up to $120 million US dollars per incident. The agreement applies to facilities within the jurisdiction of EU Member States, Norway, the Isle of Man, and the Faroe Islands. OPOL defines pollution damage as a direct loss or damage by contamination resulting from a discharge of oil. Remedial measures mean reasonable measures to prevent, mitigate or eliminate pollution damage or to remove or neutralise the oil.[28] Accordingly purely environmental damage through the diminution of the value of natural resources is not covered by the agreement. It may also be questioned what a direct loss means in a concrete case. Surely a fisherman fishing near polluted waters can claim for property loss if the slick overtakes his nets before he can retrieve them. But it will be difficult to recover an economic loss as a result of lower fish production in a polluted area. It will be even more difficult if a loss results from marketing problems due to people's fear that the catch may be contaminated.[29]

VII. Lugano Convention

The Lugano Convention,[30] adopted by the Council of Europe in 1993, aims at ensuring adequate compensation for damage resulting from activities dangerous to the environment and also provides for means of prevention and reinstatement.[31]

---

27 See *Gehring/Jachtenfuchs*, Haftung und Umwelt – Interessenkonflikte im internationalen Weltraum-, Atom- und Seerecht (1988) 195; *Gündling*, Ölunfälle bei der Ausbeutung des Festlandsockels – Zur Verschmutzung der Meere und ihrer völkerrechtlichen Kontrolle: ZaöRV 1977, 563; *UNEP* (Note 1 supra) 43.

28 The agreement is available at <www.opol.org.uk>; for further details see *Swan*, Ocean Oil and Gas Drilling and the Law (1979) 181 et seq.; *Gehring/ Jachtenfuchs* (Note 27 supra); *Bussek* (Note 2 supra) 75 et seq.

29 See for these examples *Swan* (Note 24 supra) 344.

30 Convention on Civil Liability for Damage resulting from Activities Dangerous to the Environment, Council of Europe, European Treaty Series No. 150; cf. White Paper (Note 5 supra) 52.

31 See *UNEP* (Note 1 supra) 44 et seq.; White Paper (Note 5 supra) 52.

This convention applies to the marine environment as the term "environment" includes both abiotic and biotic natural resources such as air, water, soil, fauna and flora and any interaction between these areas.[32] The term *damage* covers damage to persons and property as well as damage by impairment of the environment including the costs of measures of reinstatement and of preventive measures.[33] Measures of reinstatement are defined as any reasonable measures aiming to reinstate or restore damaged or destroyed components of the environment, or to introduce, where reasonable, the equivalent of these components into the environment.[34] This definition is quite remarkable as it does not only cover clean-up measures, but also addresses the question of what to do if the original status cannot be restored by including the costs for introducing equivalent components. By this definition, the convention paves the way for compensation of damage to the environment per se.

The compensation system outlined in the Lugano Convention is restricted to damage caused by a professionally performed dangerous activity. These activities, which are defined in detail,[35] include the production, handling, storage, use or discharge of specific dangerous substances, genetically modified organisms and microorganisms, and the operation of specific installations or sites for the treatment of wastes. Only stationary activities are covered under the scheme, excluding the transportation of substances. The operator of an activity, i.e., the person exercising control of a dangerous activity,[36] is strictly liable and is required to maintain insurance.[37] If an incident consists of a series of occurrences having the same origin the operators of any such occurrence shall be jointly and severally liable.[38] In principle, this compensation scheme could serve as a basis for claims resulting from land-based marine pollution to the extent the pollution is caused by a dangerous activity covered by the convention.[39] A remarkable aspect of this agreement is the additional right it grants to environmental organisations. They may not only request that unlawful and threatening, dangerous activities are prohibited, but also that the operator is ordered to take measures of reinstatement.[40] The Lugano Convention, however, is not yet in force.

---

32 Article 2(10).

33 Article 2(7).

34 Article 2(8).

35 Article 2(1)–(4), Annexes I and II.

36 Article 2(5).

37 Article 6, 7, 12.

38 Article 11.

39 The Convention is applicable when the incident occurs in the territory of a Party, regardless of where the damage is suffered. It also applies to incidents occurring outside the territory when the conflict of laws rules lead to the application of the law in force for the territory, cf. Article 3.

40 Article 18(1).

VIII. The Basel Protocol

Although the Basel Convention dealing with the transboundary movement of hazardous wastes and other wastes is not a maritime convention, it does cover the transport of wastes by ships, and is thus relevant to the protection of the marine environment. The Convention required parties to set out compensation regulations, and after lengthy deliberations, the Fifth Conference of the Parties to the Basel Convention in 1999 adopted the Basel Protocol.[41] This protocol will enter into force after ratification by 20 States.[42] It will apply to incidents of damage that occur during the transboundary movement of hazardous wastes and other wastes, and their disposal.[43] The term damage is defined to include the loss of life or personal injury, the loss of property or of income directly deriving from an economic interest in any use of the environment. It also covers costs of measures to reinstate the impaired environment and the costs of preventive measures.[44] Measures of reinstatement include all reasonable measures aiming to assess, reinstate or restore damaged or destroyed components of the environment.[45] While the regulation outlined in the Basel Protocol extends beyond the traditional definition of damage, it still only covers costs for measures actually taken; compensation for damaging the environment per se is not possible.

The Basel Protocol introduces strict liability for the notifier of hazardous waste movement, i e., the State of export or the exporter or generator of wastes. Other persons are liable only in case of fault.[46] Strict liability is financially limited, corresponding to the units of shipment in tonnes,[47] and must be covered by insurance.[48] The scope of application of the agreement includes the territorial sea, the exclusive economic zone and the high sea, as the Protocol applies to damage suffered in areas under national jurisdiction of a Contracting Party as well as in areas beyond national jurisdiction.[49] Where there is overlap with the HNS Convention, precedence will be given to the maritime convention,[50] provided it enters into force.

---

41 Protocol on Liability and Compensation for Damage resulting from Transboundary Movements of Hazardous Wastes and their Disposal, available from <www.basel.int/pub/protocol/html>.

42 Article 29.

43 Article 3.

44 Article 2(2)(c).

45 Article 2(2)(d).

46 Article 4, 5, see also Article 6 Basel Convention.

47 Article 12 and Annex B.

48 Article 14.

49 Article 3(3).

50 Article 11.

IX. EC Directive on Environmental Liability

Confronted with cases of severe damage to the environment resulting from human activity, the European Union saw the need to establish a framework whereby environmental damage would be prevented or remedied. Although this could have been achieved by acceding to the Lugano Convention, for various reasons the EU preferred to introduce its own legislation.[51] As a consequence, the European Parliament and the Council of the European Union adopted the Directive on environmental liability concerning the prevention and remedying of environmental damage.[52] This directive does not cover the compensation of traditional types of damage, such as health or property of persons or their economic loss,[53] but deals only with environmental damage. It promotes the prevention and remedy of environmental damage using the polluter-pays-principle. Thus, an operator whose activity has caused an environmental damage or the imminent threat of such damage is to be held financially liable.[54]

The EC Directive on Environmental Liability defines *damage* as "a measurable adverse change in a natural resource or measurable impairment of a natural resource service, which may occur directly or indirectly."[55] Damage to the marine environment covers damage to protected species and natural habitats, including any damage having significant adverse effects on reaching or maintaining the favourable conservation status of such habitats or species.[56] *Water damage* is any damage that significantly adversely affects the ecological, chemical or quantitative status or ecological potential of the waters concerned.[57] However, the area of application of the Directive with regard to the marine environment remains ambiguous. Regarding water quality reference is made to the Water Framework Directive[58] that is to be applied in coastal waters. Coastal waters are defined as the waters up to the distance of one nautical mile of the seaward side of the baseline except with respect to the chemical status , which includes all territorial waters.[59] Consequently, one must differentiate between whether the damage affects the ecological or the chemical status of the water. This distinction may be difficult to

---

51 For details see White Paper (Note 5 supra) 25 et seq.

52 Directive 2004/35/EC of 21 April 2004, OJ 2004 L 143, p. 56; for details see *Becker*, Einführung in die Richtlinie über Umwelthaftung zur Vermeidung und Sanierung von Umweltschäden: NVwZ 2005, 371 et seq.

53 Preamble, paragraph 14.

54 Preamble, paragraph 1.

55 Article 2(2); this definition is in line with the proposal made by an UNEP expert group, cf. *UNEP* (Note 1 supra) 5.

56 However, the definition of damage to protected species and natural habitats does not include adverse effects resulting from expressly authorised acts, Article 2(1)(a); cf. *Becker* (Note 52 supra) 373.

57 Article 2(1).

58 Directive 2000/60/EC of the European Parliament and of the Council of 23 October 2000 establishing a framework for Community action in the field of water policy, OJ 2000 L 327, p. 1.

59 Article 2(1) and (7).

draw in practice because determining whether damage exists ultimately depends on the adverse ecological effects caused by a negative chemical status. Regardless, the Directive does not apply to the case of water quality outside the territorial sea. However, with regard to damage to species and habitats reference is made to the Council Directive on the conservation of wild birds[60] and the Council Directive on the conservation of natural habitats and of wild fauna and flora[61] which are also applicable in the exclusive economic zone of an EU Member State,[62] as far as protected areas following the requirements of the directives are designated. This difference in the scope of application is confusing and raises the question of whether the implications of the directive for the marine environment have been adequately considered.

The Directive also applies to environmental damage caused by an occupational activity.[63] Annex III of the Directive specifies this activity as one which poses potential or actual risk to the environment including the operation of specific installations, waste management, discharges into inland surface and ground water which require prior authorisation, manufacture, use, handling and transport of dangerous substances. Also covered is the release into the air of any of the respective polluting substances. For other occupational activities not listed in this Annex the directive only applies, e.g., if the operator has been at fault or negligent.[64]

The Directive requires Member States to ensure that environmental damage is remedied. This is a public duty that must be implemented by competent authorities designated by the Member States.[65] The competent authority may require the operator to take necessary preventive or remedial measures. Alternatively the competent authority may implement such measures itself, or have them implemented by a third party.[66] In all cases the operator shall bear the costs,[67] reflecting the principle that the operator is strictly liable. However, under national law, the operator may be allowed not to bear the costs when he is not at fault or negligent and the damage results from events or activities allowed by applicable laws or authorised by a permit.[68] This approach is also taken in instances where the damage has been caused by activities that were not considered harmful according to the state of scientific and technological knowledge. While the strict liability of the operator is limited in these cases, the competent authority may be required to take action in relation to the damage in question.

---

60 Directive 79/409/EEC of 2 April 1979, OJ 1979 L 103, p. 1.

61 Directive 92/43/EEC of 21 May 1992, OJ 1992 L 206, p. 7.

62 See *Brandt/Gassner*, Seeanlagenverordnung (2002) § 3 No. 59, with further references.

63 For definition see Article 2(7).

64 Article 3.

65 Article 11; cf. Preamble, paragraph 15.

66 Articles 5 – 7.

67 Article 8.

68 Article 8(4); as far as adverse effects to protected species or natural habitats are concerned the damage itself may be negated in case of authorised activities, cf. supra Note 56.

The purpose of the remedial measures[69] is to restore the damaged natural resources to baseline conditions. This also includes the compensation of interim losses of natural resources until the restoration measures have taken effect. Where the remedial measures do not result in full restoration complementary measures are to be undertaken to provide similar natural resources at the damaged or an alternative site. With the aim of enabling operators to use financial guarantees to cover their responsibilities the development of financial security instruments shall be encouraged.[70] The directive does not deal with cost allocation in cases of multiple party causation especially concerning the apportionment of liability between the producer and the user of a product, but leaves it to the Member States to introduce national regulations.[71]

Persons adversely affected or likely to be adversely affected by environmental damage may request that the competent authority take remedial action. Non-governmental organisations promoting environmental protection are also entitled to request the competent authority take action.[72]

Environmental damage can only be remedied by means of the liability mechanism if there are one or more identifiable polluters, if the damage is concrete and quantifiable, and if a causal link can be established between the damage and the identified polluter.[73] Therefore, the Directive cannot be applied to widespread, diffuse pollution where it is impossible to link negative environmental effects to acts of a certain individual actor.

Concerning marine pollution from ships, the Directive shall not apply to incidents where compensation falls within the scope of the IMO liability conventions. The Directive shall also be without prejudice to the right of the operator to limit his liability in accordance with the Convention on Limitation of Liability for Maritime Claims, 1976.[74]

X. Evaluation and Conclusions

At issue is to what extent present international legislation, including the EC-Directive, provides sufficient legal possibilities for compensating damage caused by marine pollution.[75]

---

69 Details are set out in Annex II.

70 Article 14.

71 Article 9.

72 Article 12; this regulation is in line with Article 9(2) of the Aarhus Convention.

73 Article 4(5).

74 Article 4(2) and (3).

75 In this context neither the compensation schemes for nuclear damage nor for damage from seabed mining, based on Part XI of UNCLOS, are taken into consideration; see for information de *La Fayette* (Note 5 supra) 249 ff., 253 ff.; *Bothe*, The Protection of the Marine Environment against the Impacts of Seabed Mining: An Assessment of the New Mining Code of the International Seabed Authority, in: Marine Issues – From a Scientific, Political and Legal Perspective, ed by. Ehlers/Mann-Borgese/Wolfrum (2002) 229 et seq.

Despite considerable international efforts a number of relevant deficits still exist. Regretfully, by far not all of the international regulations are in force yet. Worldwide compensation regimes exist[76] only for one source of marine pollution, i.e., shipping. Damage caused by offshore activities in the European seas may be compensated by a private industry agreement, but compensation claims are limited to $120 million US dollars. The Lugano Convention, a legal instrument developed on European level, which may be honoured as an important approach for the development of the environmental law, could serve as a basis in particular for compensation of pollution from land based sources, however it is superseded by the EC-Directive on environmental liability.

The area of application for each instrument differs. Many are restricted to the territorial sea and the exclusive economic zone,[77] and only the Basel Protocol includes the high seas. From a German perspective, this might be circumstantial as the North Sea and the Baltic Sea, being the areas of direct German interest, are mostly covered by exclusive economic zones of the bordering States. Of greater importance is that the EC-Directive due to the quite ambiguous definition of the term *water damage* at best is applicable in the territorial sea area; only in very specific cases also pollution in an exclusive economic zone may be included via the Birds and FFH Directives.[78]

The scope of compensation provided for in the international regimes differs depending on the definition of damage. Most of the international regulations stem from the traditional definition of damage, providing for restoration measures only when compensating for concrete financial losses. The introduction of complementary or equivalent components in case full restoration is not possible is only considered by the Lugano Convention and more recently by the EC-Directive, though many details are still unclear. In the German legal system, the distinction between the impairment of private property rights and duties under public law must be observed, thus, it makes sense that the EC-Directive deals with environmental damage only. It is doubtful whether the aim to compensate environmental damage per se is completely achieved; this would require a clear regulation stating that if restoration is impossible there will be compensation for the value of the permanently damaged natural resource. Where such compensation is provided for, a further issue becomes how to calculate damage that is irreparable or unquantifiable.[79]

All of the regimes discussed above make the operator or owner strictly liable, particularly when there is involvement in dangerous activities. For reasons of legal certainty and economic considerations, in most regimes a cap on the amount of liability is established in combination with the maintenance of an insurance or some other financial security. After the 'Prestige' accident there has been much important discussion regarding whether a specific amount of liability is justified

---

76 This situation is also critizised by *de La Fayette* (Note 5 supra) 232.

77 As concerns the problems related to compensation regulations on the high seas cf. *Bussek* (Note 2 supra) 110.

78 Cf. Notes 60 and 61 supra.

79 For criteria to be developed cf. *UNEP* (Note 1 supra) 117.

and sufficient. Compensation claims generally require the affected party has a direct legal interest. This is problematic in instances of pure damage to the marine environment. In this context, it is remarkable that the EC-Directive entitles non-governmental environmental organisations to request the competent authority to take action. A more general approach could be to designate public trustees to bring forth claims on behalf of the marine environment.[80]

Land based inputs, the dominant source of marine pollution, generate two environmental problems that remain unsolved by present compensation systems. Land based inputs are more or less continuous, routinely discharged pollutants. The corresponding activities are mostly based on specific permits granted by the responsible authorities or are caused by normal operations, allowed under national law, where all of the given limit values and environmental protection requirements are met.[81] As an example the input of nutrients may be mentioned, which cannot be considered as a dangerous activity, but nonetheless nutrients represent one of the main pollution problems. This problem of lawful interferences is accounted for in the Lugano Convention by exempting damages caused by pollution at tolerable levels under local relevant circumstances,[82] and also in the EC Directive with regard to adverse effects for protected species and habitats.[83] For the rest the EC-Directive follows a similar approach by authorising Member States to allow the operator not to bear the costs thereby modifying its concept of strict liability.

Furthermore these inputs lead to what may be called a "creeping" pollution, which is not caused by one kind of activity, but is the result of the summation of different activities and inputs. In cases of creeping pollution, no individual polluter can be identified and no concrete causal link between an individual activity and concrete damage can be established. Only when individual causation can be proven is compensation possible under the present regulations, including the EC-Directive. Such causation can be shown, for example, in the case of a specific incident or where the pollution is in close proximity to a plant where dangerous substances are discharged into the water. Apart from such incidents or where dangerous activities result in local effects marine pollution is often widespread and diffuse in character, and is caused by a multitude of activities and factors.

Bearing in mind that neither shipping, offshore activities, nor occasional onshore incidents are the main causes of marine pollution, it becomes evident that the existing compensation regimes are not sufficient and appropriate for preventing and remedying environmental damage from marine pollution. With regard to land based pollution, additional legal instruments are required to cover the costs of restoration measures in the rare cases where restoration seems possible, and most importantly, the costs of introducing complementary and equivalent components.

---

80 Cf. *UNEP* (Note 1 supra) 110 et seq.

81 Rightly *UNEP* (Note 1 supra) 103, points out that this represents a major gap in liability agreements.

82 Article 8 (d) of the Lugano Convention.

83 Cf. Note 56 supra.

Where a permit has been granted to allow for certain imputs, a compensation scheme comparable to the German law system is conceivable. Under the German Federal Water Act a permit or licence may be granted subject to the imposition of certain conditions such as measures necessary to compensate for impairments of the ecological or chemical status of the water body, and obligations imposed on the operator to contribute to the costs, if the measures are taken by a public corporation.[84]

However, the German compensation scheme outlined above does not solve the problem of combined pollution damage as a summation from diffuse and long-range inputs. In these cases a fund system could be considered. The fund established under the Fund Convention[85] might serve as a starting point,[86] as it could be used to compensate for marine pollution damage that cannot be linked to specific activities. It could be financed through fees levied for onshore activities that deliberately or accidentally introduce pollutants into the sea.[87] Fees could function as an economic instrument that provides an incentive to reduce inputs. The fund should also compensate deterioration of the marine environment by contributing to general costs for marine protection measures, even if concrete restoration measures are not possible. Examples of such general costs include measures to introduce any equivalent components and preventive measures for preparedness.[88] Since protection of the marine environment cannot be solely achieved on a national level, a regional, or preferably, a European fund system should be developed. This could be an important step for States to meet their obligations to protect the marine environment under UNCLOS.

---

84 Paragraph 4 subparagraph 2 no. 2 a and 3 Federal Water Act, BGBl. 2002 I p. 3167; cf. *Hoffmeister/Kokott*, Öffentlich-rechtlicher Ausgleich für Umweltschäden in Deutschland und in hoheitsfreien Räumen, Berichte 9/02 des Umweltbundesamtes, p. 91.

85 Cf. Note 2 supra 3.

86 *UNEP* (Note 1 supra) 117, supports the idea of a fund, established by either State or industry as a method of collective reparation in case that restoration measures are not technically feasable or reasonable.

87 At least in principle such a fee system would be comparable to the German Federal Water Fee Act; cf. *Hoffmeister/Kokott* (Note 77 supra) 93.

88 See Bussek (Note 2 supra) 106 et seq., 122, who proposes a "blanket allowance" for the preservation of the seas.

Maritime Pollution – Compensation or Enforcement?

Rüdiger Wolfrum

I. Introduction

The *United Nations Convention on the Law of the Sea* and other international agreements, in particular those concerning the protection of the marine environment against oil pollution, provide for an individual liability of polluters. There has been an increasing trend to hold individuals or corporations liable for oil pollution damage. No such trend exists in respect of State liability, neither subsidiary in cases where the individual polluter does not provide compensation, nor in cases where environmental damage results from the violation of a State's own non-compliance of international environmental obligations. In respect of the latter general international law on state responsibility is of relevance.[1]

Liability for oil pollution damage can be seen from two different perspectives. It may be considered as a means to enforce environmental standards or – at least – to supplement existing enforcement mechanisms. It may, however, also be seen to balance the various economic interests concerning the use of maritime space. Oil pollution damage may be detrimental to tourist activities, to activities concerning the exploitation of living resources (e.g., fishing or fish farming), or to marine scientific research. In general, the balancing approach seems to prevail. It is for that reason that the existing liability regimes mostly do not provide for the payment of compensation for purely environmental damages, i.e., damages that do not result in economic loss or damage to property. The fact that existing liability regimes concentrate on compensation for damage to property or economic loss does not mean that they do not have the effect of enforcing environmental standards, though. Such indirect effect exists and it is one of the objectives pursued by the regimes in question.

---

1 For further details see *Dahm/Delbrück/Wolfrum*, Völkerrecht, Vol. I/3 (2nd ed. 2002) 873 et seq.

II. Existing Regimes

1. Convention on the Law of the Sea

The *United Nations Convention on the Law of the Sea* (Convention) provides, in article 235, paragraph 1, that State Parties are liable for the protection of the marine environment, in accordance with public international law. State Parties are obligated to provide legal remedies in their national laws for the compensation of damage to the marine environment caused by natural persons or entities within their jurisdiction (article 235, paragraph 2). In article 235, paragraph 3, the Convention lays down certain specifications, albeit vague, for the further development of an international liability regime concerning the protection of the marine environment.

According to article 232, States Parties are liable for damage resulting from the measures they take to enforce environmental provisions, where such measures are unlawful or exceed those reasonably required in the light of available information. Article 263, paragraph 3, of the Convention is also significant, as it refers to the liability rule of article 235 for damages "caused by pollution of the marine environment arising out of marine scientific research undertaken by them or on their behalf."

These provisions are rather embryonic in nature, and require further development. Therefore, it would be an overstatement to say that the Convention provides for a liability regime that covers either individual or State liability for environmental damage. In assessing the rules, it is important to consider that the Convention was adopted in 1982, and thus, constituted a progressive approach for the early eighties. For that reason, it is not surprising that the liability mechanism does not play a central role in the implementation of environmental standards.

2. International Convention on Civil Liability for Oil Pollution Damage and the Protocol of 1992

The following instruments have to be mentioned in respect of liability for oil pollution damage, namely: the *International Convention relating to Intervention on the High Seas in Cases of Oil Pollution Casualties*, the *International Convention on Civil Liability for Oil Pollution Damage*, the *Oil Liability Convention*, and the *1971 International Convention on the Establishment of an International Fund for Oil Pollution Damage* (*1971 Fund Convention*).

Of special relevance in the development of the *Oil Liability Convention* and the *Fund Conventions* were voluntary agreements between oil companies, such as a tanker owners' voluntary agreement concerning liability for oil pollution, the supplement to tankers owners' voluntary agreement concerning liability for oil pollution and the Contract Regarding an (Interim) Supplement to Tanker Liability for Oil Pollution.[2]

---

[2] The agreements are reproduced in: *Gehring/Jachtenfuchs*, Haftung und Umwelt, (1989) 267 et seq.; for details see *Bussek*, Schutz der Meere vor Verschmutzung,

The *Civil Liability Convention for Oil Pollution Damage* was designed in reaction to the dangers of pollution posed by the world wide maritime carriage of oil in bulk. Its main objectives are to ensure that adequate compensation is available to persons who suffered damage by spills of persistent oil from tankers, and to harmonize international rules and procedures for determining questions of liability and for providing adequate compensation in such cases.[3] The liability and compensation regime created by that Convention is based on strict liability of the polluter (i.e., the ship owner transporting oil).[4] The owner of a seagoing vessel that carries oil as a bulk cargo is liable, regardless of fault, for pollution damage caused by oil escaping from or released from the vessel.

The framework for the liability regime was originally established by the *1969 International Convention on Civil Liability for Oil Pollution Damage* (*1969 Civil Liability Convention*) and the *1971 Fund Convention.* These agreements were amended by two Protocols of 1992, which entered into force on 30 May 1996.[5] The entering into force of these Protocols means that a comprehensive system for the settlement of tanker incidents is available worldwide. The flag State of the tanker and the nationality of the shipowner are irrelevant for determining the scope of application. The private agreements referred to above have, therefore, lost their reason for existence and accordingly were discontinued in February 1997.

Under the 1992 amending Protocol, the Convention not only covers vessels that carry oil as bulk cargo, but also all vessels that are designed to carry oil as a mass product. This includes empty tankers that cause pollution damage, for example, as a result of prohibited tank cleaning or bunker oil. The Convention does not cover ships that carry gasoline for their own use. Liability for the accidental release of such gasoline is covered by the ordinary liability of ships' operators. The Convention does not only cover accidental damage; at issue is whether the damage is caused by an environmentally relevant event (escape or discharge of oil) within the meaning of this Convention.

(1993) 666 et seq.; *Bornheim*, Haftung für Grenzüberschreitende Umweltbeeinträchtigung im Völkerrecht und Internationalem Privatrecht (1995) 96 et seq. For a more detailed description of the voluntary liability agreements for compensation of oil pollution damage see *Wolfrum/Langenfeld*, Environmental Protection by Means of International Liability Law, Berichte 6/99 (1999) 20 et seq.; *Langenfeld/Minnerop*, Environmental Liability Provisions in International Law, in: Environmental Liability in International Law – Towards a Coherent Conception, ed. by *Wolfrum/Langenfeld/Minnerop*, Berichte 2/05 (2005) 3.

3 *Langenfeld/Minnerop* (Note 2 supra) 5 et seq.

4 For a comparison between the international and the US regimes regulating oil pollution liability compensation see *Noqinho Kim*, Marine Policy 27 (2003) 265 et seq. The owner of a seagoing vessel that carries oil as a bulk cargo is liable, regardless of fault, for pollution damage caused by oil escaping from or released from the vessel.

5 As of 1 September 2006, 113 States had ratified the 1992 Civil Liability Convention, and 98 States had ratified the 1992 Fund Convention.

The *1992 Civil Liability Convention* extended the scope of persons against whom claims for compensation may be brought. It states *inter alia* that persons who perform salvage or assistance can be held liable only in cases where the damage resulted from their personal act or omission. Initially, the objective of the rules was to channel the entire liability to the owner of the vessel.

There is no provision for subsidiary State liability in the event that the ship owner does not pay compensation for the damage caused. The regime's approach to the management of the resources of the deep seabed is different in this respect. Article 139, paragraph 2, of the Convention places liability on State parties which have not taken all necessary and appropriate measures to secure effective compliance by the operators with the respective obligations of the latter which it has sponsored. Technically, this makes a State liable for the violation of its own international obligations rather than for private activities.[6] This was one of the requests of the United States of America, whereas the then Soviet Union was in favour of a liability of States for violations committed by private entities directly. In respect of oil pollution damage there is no mentioning of State responsibility in the respective agreements.

According to article 3, paragraph 1, of the *1992 Civil Liability Convention*, the owner of the ship at the time of the incident shall be liable for any pollution damage caused by oil that has escaped or has been discharged from a ship as a result of the incident, if none of the exceptions apply.[7] Compared to the *1969 Civil Liability Convention* this central provision has undergone some modification. Under article 1, paragraph 6, of the Convention of 1969, pollution damage had been defined as "loss or damage caused outside the ship carrying oil by contamination resulting from the escape or discharge of oil from the ship, wherever such escape or discharge may occur, and includes the costs of preventive measures and further loss or damage caused by preventive measures." The term "incident" meant any occurrence or series of occurrences having the same origin, which causes pollution damage. According to paragraph 2 of paragraph 6 of the *1992 Civil Liability Convention*, pollution damage now means: loss or damage caused outside the ship by contamination resulting from the escape or discharge of oil

---

6 This issue was highly disputed. It is a question whether article 139, para. 2, of the Convention on the Law of the Sea in connection with article 4 of Annex III of the Convention of the Law of the Sea constitutes a new responsibility of States, or whether it is nothing more than the reconfirmation of State responsibility under general international law. The first words of Article 139(2) of the Convention of the Law of the Sea seem to confirm the latter interpretation.

7 Article 3(2) provides for several exceptions to owner liability. These include an act of war, hostilities, civil war, insurrection, or a natural phenomenon of an exceptional, inevitable and irresistible character; damage wholly caused by an act or omission done with the intent to cause damage by a third party; or damage wholly caused by negligence or other wrongful action by any government or other authority responsible for the maintenance of lights or other navigational aids in the exercise of that function. Another exception is that the damage resulted from an act or omission with intent to cause damage by the person who suffered the damage or from the negligence of that person.

from the ship, wherever such escape or discharge may occur, provided that compensation for impairment of the environment, other than loss of profit from such impairment, shall be limited to costs of reasonable measures for reinstatement actually undertaken or to be undertaken, the costs of preventive measures and further loss of damage caused by preventive measures.

The 1992 amendment introduces two policies. It limits the potential amount of damage costs incurred by referring to reasonable measures to reinstate the contaminated environment. This excludes the obligation to compensate environmental damages where reinstatement measures are, for one or other reason, excluded. This does not mean, however, that purely environmental damages are excluded from being covered under this regime.[8]

The 1992 Protocol widens the territorial scope of the *Civil Liability Convention* compared with its 1969 version to cover pollution damage not only in the internal and coastal waters but also in the exclusive economic zone or equivalent areas of a State Party. Thus, only pollution damage on the high seas does not come within the scope of the *1992 Civil Liability Convention.*

As already indicated, compensation may even include the costs of preventive measures if there was a grave and imminent threat of pollution damage: according to article 1, paragraph 8, of the *1992 Civil Liability Convention*, the definition of "incident" includes any occurrence, or series of occurrences, with the same origin, which causes pollution damage or creates a grave and imminent threat of causing such damage. This means expenses are recoverable even in cases where no oil spill occurred, but there was a threat of such pollution damage.

In the past, interpretation of the term "pollution damage" has been the subject of considerable controversy between the Fund and States Parties. Meanwhile, a consensus between the States Parties has been reached on essential aspects of the scope of compensation.

It is undisputed that compensation is payable in full for the cost of a complete clean-up of polluted shores, river banks and beaches, beach facilities, ships and marine facilities. Consultations between the shipowner concerned, his liability insurers, the Fund (if necessary), and the State authorities concerned are regularly held before, and in parallel with, the clean-up measures, which are the responsibility of the competent State administrations.

Compensation for damages is still subject to dispute, particularly with respect to reimbursement of fixed costs, and also to compensation for lost profits in cases not involving property damage, distinguishing damage prevention from salvage costs and compensating for residual ecological damage.

Some words are necessary concerning fixed costs. Many States keep human and natural resources constantly available for cleaning up oil pollution damage. Public service personnel are also regularly employed for clean-up work. It is disputed whether such costs are to be borne by the shipowner having caused the damage. There is no doubt that the additional cost occasioned by the relevant

---

8 On this subject (although dealing with European Law), see *Maes*, Estimating Damages under the 2004 EC Directive on Environmental Liability, in: Marine Damage Assessment, ed. by Maes (2005) 3 et seq.

measures are to be reimbursed, but it is not yet clear whether the normal pay for personnel which will be used in the case of an accident will be covered. The Fund's settlement practice is basically to refund fixed costs only up to the appropriate limit. On this basis, costs for providing equipment and personnel are calculated for the duration of the measures in question.

The *1969 Civil Liability Convention* left open the question whether loss of profits by hotels and restaurants along oil-polluted beaches owing to the absence of tourists was to be considered as pollution damage. In its settlement practice, the Fund included such losses within the scope of the term pollution damage at an early stage. On this basis, for example, it compensates for losses incurred by enterprises whose existence is based on a justified reliance on the intactness of the marine environment. This broad interpretation of pollution damage benefits the tourism sector, particularly to the extent that the pollution of beaches directly results in a decline of tourism.

In the Fund's practice a distinction between damage prevention costs and salvage costs has always played a major role, partly because claims of substantial size had to be assessed. In effect, the distinction has to be based on the subjective criteria, which gives rise to special problems. Thus, if the purpose of a rescue measure is primarily to save the ship and its cargo, the resulting costs are salvage costs, which are to be borne by the hull insurer and the cargo insurer. If, on the other hand, the primary purpose of the measure is to prevent further oil pollution, then it is a damage prevention measure, which is to be borne by the shipowner's liability insurer and/or the Fund.

The 1992 Protocol raised the liability limits. The prerequisite for limitation of shipowners' liability is that they should establish a fund for the total sum representing the limit of their liability with a court or other competent authority in any one of the States Parties. Until adequate coverage has been provided for, the maximum amount of liability, the owner's liability, is unlimited as to this amount. Finally, the limitation of liability may not be invoked if the incident occurred as a result of intent or gross negligence on the part of the shipowner.

To ensure compliance with the liability provisions of the *1992 Civil Liability Convention*, article 7, paragraph 1, requires that the owner of a ship carrying more than 2 000 tonnes of oil in bulk as cargo, shall maintain insurance or other financial security to cover the liability under the Convention. A certificate to this effect is to be issued by the competent authority of the flag State under article 7, paragraph 2, and under article 7, paragraph 4, of the Convention.

The regime on civil liability as established by the 1992 Protocol was supplemented by the *International Convention on the Establishment of an International Fund for Compensation of Oil Pollution Damage*. The main objectives of the Fund Convention are to provide compensation for pollution damage to the extent that the protection afforded by the *1992 Civil Liability Convention* is inadequate, and to give relief to shipowners in respect of the additional financial burden imposed upon them by the *Civil Liability Convention*. The three main cases where the Fund will have to intervene are the shipowner is exempt from liability, or he is incapable of meeting his financial obligation or the damage exceeds the limits.

On 18 October 2000, further amendments to the 1992 Protocol were adopted. They raised the compensation limits by 50 percent compared to the limits set in the 1992 Protocol, and entered into force on 1 November 2003 under tacit acceptance.

During a diplomatic conference held in London on 16 May 2003, a Protocol establishing an International Oil Pollution Compensation Supplementary Fund was adopted. The objective of the Protocol is to supplement the compensation available under the *1992 Civil Liability* and the *Fund Convention* with an additional third tier of compensation. The Protocol is optional.[9]

3. Assessment

An overall assessment of the liability regime must be positive. It has established a framework which guarantees that environmental, as well as economic damages resulting from oil spill from tankers are covered, and respective restoration is undertaken. The first call on the Fund was made in 1979, when the Soviet tanker *Antonio Gramsci* ran aground off the Baltic coast. Already by the end of 1996, the Fund had been called on in 75 cases. This does not mean that preventive measures have become unnecessary. Contrary preventive measures may be a more effective means of protection than recovery activities.

The *1992 Civil Liability Convention* has a channelling effect. Claims for pollution damage under the *1992 Civil Liability Convention* can be made only against the shipowner, and not against servants or agents of the owner, members of the crew, the pilot, the charterer or the manager. This does not prevent the shipowner from having recourse to such persons.

Nevertheless, this regime can still be improved. One of the most controversial questions remaining is compensation for residual ecological damage, i.e., damage that cannot be eliminated by reinstatement measures and the value of which cannot be calculated on the basis of traditional parameters.

The question of including residual ecological damage in the obligation to make compensation arises not only with regard to oil pollution damage. It is one of the most difficult, and from an economic point of view, possibly one of the most significant problems. Particularly in view of the problematic nature of compensation for ecological damage, the new definition of pollution in the 1992 Protocols is confined to compensation for appropriate reinstatement measures. This is in line with the previous practice of the Fund, which has hitherto refused compensation for purely ecological damage. The issue became acute in connection with a tanker incident in the Baltic, when the USSR claimed compensation for loss calculated under the statutory formula for residual water pollution.

In recent years, national courts have often addressed the issue of the extent to which compensation is payable for residual ecological damage. In this context, the

---

9 The International Group of P&I Clubs has introduced, on a voluntary basis, a compensation package consisting of two agreements, the Small Tanker Pollution Indemnification Agreement and the Tanker Oil Pollution Indemnification Agreement. Both entered into force in February 2006.

Patmos case has become quite important. In 1985, the Greek tanker *Patmos* collided with a Spanish tanker off the coast of Italy, polluting the Sicilian coast. The Italian authorities took extensive preventive measures. The Italian government subsequently filed a claim for compensation amounting to 2.3 million British pounds for ecological damage to the marine environment. The Fund rejected the claim, after the Court of First Instance in Messina had also rejected the claim. The Italian authorities won their case in the Court of Appeal. The Court of Appeal decided that the owner of the tanker *Patmos* had to pay a sum of 830 000 British pounds to the Italian government for ecological damage.[10] In its reasoning, the Court argued that the concept of damage in the *Liability Convention* was sufficiently broad to cover residual ecological damage. The *Patmos* case took place before the 1992 Protocol entered into force, thus, compensation for residual ecological damage under the liability regime established by the *Liability Convention* and the *Fund Convention* is no longer possible.

However, even the more precisely worded 1992 definition of damage gives reasons to expect problems in the practical calculation of the compensation payable. If the compensation would move in this direction, it would shift the regime from a primarily compensatory into an enforcement regime.

---

[10] Messina Appeals Court 24 December 1993, Summary of the Judgement in: RECIEL 1995, 341-2. See also: *Maffay*, Compensation for Ecological Damage in the Patmos Case, in: International Responsibility for Environmental Harm, ed. by Francioni/Skowatsi (1991) 381 et seq.

The International Oil Pollution Compensation Funds and the International Regime of Compensation for Oil Pollution Damage

Måns Jacobsson

I. Introduction

The international regime for the compensation of pollution damage caused by oil spills from tankers is based on two treaties adopted under the auspices of the International Maritime Organization (IMO), the 1992 International Convention on Civil Liability for Oil Pollution Damage (1992 Civil Liability Convention), and the 1992 International Convention on the Establishment of an International Fund for Compensation for Oil Pollution Damage (1992 Fund Convention). These Conventions replace two corresponding Conventions adopted in 1969 and 1971 respectively.

The 1992 Civil Liability Convention governs the liability of shipowners for oil pollution damage. The Convention lays down the principle of strict liability for shipowners and creates a system of compulsory liability insurance. Shipowners are normally entitled to limit their liability to an amount which is linked to the tonnage of his ship.

The 1992 Fund Convention, which is supplementary to the 1992 Civil Liability Convention, set up an intergovernmental organisation, the International Oil Pollution Compensation Fund (1992 Fund), which provides additional compensation to victims when compensation under the Civil Liability Convention is inadequate. By becoming party to the Fund Convention, a State becomes a member of the 1992 Fund. The Organisation has its headquarters in London.

The 1992 Fund succeeds a previous organisation, the 1971 Fund, which is at present being wound up.

On 15 April 2005, 106 States were parties to the 1992 Civil Liability Convention, and 93 States were parties to the 1992 Fund Convention.

A third tier of compensation in the form of a Supplementary Fund was established on 3 March 2005 by means of a Protocol adopted in 2003. To date nine States have ratified the Protocol.

The States which are parties to the 1992 Conventions and the Supplementary Fund Protocol are listed in the Annex.[1]

II. Main Features of the 1992 Conventions

The 1992 Conventions apply to pollution damage suffered in the territory (including the territorial sea) and the exclusive economic zone (EEZ) or equivalent area of a State party to the respective Conventions. " 'Pollution damage' is defined in the 1992 Conventions as damage caused by contamination and includes the cost of 'preventive measures', i.e. measures to prevent or minimise pollution damage."

The 1992 Conventions apply to ships which actually carry oil in bulk as cargo, i.e. generally laden tankers, as well as to spills of bunker oil from unladen tankers in certain circumstances.

Liability for pollution damage rests on the registered owner of the ship from which the oil originated. Shipowners have strict liability for pollution damage (with very limited defences) and are obliged to cover their liability by insurance. Shipowners are normally entitled to limit their liability to an amount which is calculated on the basis of the tonnage of the ship, and which – after increases by some 50% with effect from 1 November 2003 – ranges from US $7 million for small ships to US $136 million for large tankers.[2]

Shipowners are deprived of the right to limit their liability if it is proved that the pollution damage resulted from the shipowner's personal act or omission, committed with the intent to cause such damage, or recklessly and with knowledge that such damage would probably result.

Claims for pollution damage under the 1992 Civil Liability Convention can be made only against the registered owner of the ship concerned. This does not preclude victims from claiming compensation outside the Convention from persons other than the owner. However, the Convention prohibits claims against the servants or agents of the owner, the crew, the pilot, the charterer (including a bareboat charterer), manager or operator of the ship, or any person carrying out salvage operations or taking preventive measures.

The compensation payable by the 1992 Fund in respect of an incident is limited to an aggregate amount which, with effect from 1 November 2003, was increased from US $205 million to US $310 million, including the sum actually paid by the shipowner (or his insurer) under the 1992 Civil Liability Convention.

The 1992 Fund is financed by contributions levied on any entity which has received in one calendar year more than 150 000 tonnes of crude or heavy fuel oil

---

1 Information on the international compensation regime and the 1992 Fund and the Supplementary Fund is available on the Funds' web site at:<http://www. iopcfund. org>.

2 The unit of currency in the 1992 Conventions is the Special Drawing Right (SDR) as defined by the International Monetary Fund. In this document the SDR has been converted into US dollars at the rate applicable on 15 April 2005, ie 1 SDR = US $1.502910.

(contributing oil) in a State party to the 1992 Fund Convention after sea transport. Member States are obliged to submit annually to the Fund reports on the quantities of contributing oil received.

The Japanese oil industry is the major contributor to the 1992 Fund, paying 18% of the total contributions. The Italian oil industry is the second largest contributor paying 10%, followed by the oil industries in the Republic of Korea (9%), the Netherlands (8%), France (7%), India (7%), United Kingdom (5%), Singapore (5%) and Spain (5%).

The 1992 Fund has an Assembly, which is composed of representatives of all 1992 Fund Member States. The Assembly is the supreme organ governing the 1992 Fund, and it holds regular sessions once a year. The Supplementary Fund has its own Assembly composed of representatives of its Member States.

The 1992 Fund, the 1971 Fund and the Supplementary Fund have a joint Secretariat. The Secretariat is headed by a Director and has at present 27 staff members.

The Director has been granted extensive authority to approve claims for compensation.

III. Claims Settlement

1. Claims experience

Since their establishment, the 1971 and 1992 Funds have been involved in approximately 135 incidents and have made compensation payments totalling some US$860 million. The Supplementary Fund has thus far not been involved in any incidents.

In the great majority of these incidents, all claims have been settled out of court. To date, court actions against the Funds have been taken in respect of only a handful of incidents.

The cases involving the largest total payments are as follows:

Incident	*Payments to claimants*
Antonio Gramsci (Sweden, 1979)	US $ 17 million
Tanio (France, 1986)	US $ 36 million
Haven (Italy, 1991)	US $ 57 million
Aegean Sea (Spain, 1992)	US $ 64 million
Braer (United Kingdom, 1993)	US $ 86 million
Keumdong Nº 5 (Republic of Korea, 1993)	US $ 21 million
Sea Prince (Republic of Korea, 1995)	US $ 40 million
Yuil Nº 1 (Republic of Korea, 1995)	US $ 30 million
Sea Empress (United Kingdom, 1996)	US $ 59 million
Nakhodka (Japan, 1997)	US $209 million
Nissos Amorgos (Venezuela, 1997)	US $ 21 million
Erika (France, 1999) (payments up to 15-4-05)	US $106 million
Prestige (Spain, France, Portugal, 2002) (payments up to 15-4-05)	US $ 75 million

A major oil spill can give rise to a large number of claims. The *Erika* incident resulted in over 6 900 compensation claims, of which over 50% were presented by businesses in the tourism sector and 27% originated from the fishery and mariculture sectors.

2. Admissibility of claims for compensation

The 1992 Fund can pay compensation to a claimant only to the extent that the claim meets the criteria laid down in the 1992 Fund Convention.

The Funds have acquired considerable experience with regard to the admissibility of claims. In connection with the settlement of claims they have developed certain principles regarding the meaning of the definition of "pollution damage" which is specified as "damage caused by contamination".

The 1992 Fund has published a Claims Manual which contains general information on how claims should be presented and sets out the general criteria for the admissibility of various types of claims. A revised version of the Claims Manual adopted by the Assembly was published in May 2005.

Decisions on the admissibility of claims which are of general interest are reported in the Funds' Annual Report.

Some of the main types of claim are dealt with below.

a) Property damage

Pollution incidents often result in damage to property: the oil may contaminate fishing boats, fishing gear, yachts, beaches, piers and embankments. The 1992 Fund accepts costs for cleaning polluted property. If the polluted property (e.g. fishing gear) cannot be cleaned, the Fund compensates the cost of replacement, subject to deduction for wear and tear. Measures taken to combat an oil spill may cause damage to roads, piers and embankments and thus necessitate repair work, and reasonable costs for such repairs are accepted by the Fund.

b) Clean-up operations on shore and at sea, and preventive measures

The 1992 Fund pays compensation for expenses incurred for clean-up operations at sea or on the shore. Operations at sea may relate to the deployment of vessels, the salaries of crew, the use of booms and the spraying of dispersants. In respect of onshore clean-up, the operations may result in major costs for personnel, equipment, absorbents etc.

Claims for measures to prevent or minimise pollution damage are assessed on the basis of objective criteria. The fact that a government or other public body decides to take certain measures does not in itself mean that the measures are reasonable for the purpose of the Conventions. The technical reasonableness is assessed on the basis of the facts available at the time of the decision to take the measures. However, those in charge of the operations should continually reappraise their decisions in the light of developments and further technical advice.

Claims for costs are not accepted when it could have been foreseen that the measures taken would be ineffective. On the other hand, the fact that the measures prove to be ineffective is not in itself a reason to reject a claim for the costs incurred. The costs incurred, and the relationship between those costs and the benefits derived or expected, should be reasonable. In the assessment, the 1992 Fund takes account of the particular circumstances of the incident.

Measures taken to prevent or minimise pollution damage (preventive measures) are compensated by the 1992 Fund. Measures may have to be taken to prevent oil which has escaped from a ship from reaching the coast, eg by placing booms along the coast which is threatened. Dispersants may be used at sea to combat the oil. Costs for such operations are in principle considered as costs of preventive measures. It must be emphasised, however, that the definition only covers costs of *reasonable* measures.

c) Consequential loss and pure economic loss

The 1992 Fund accepts in principle claims relating to loss of earnings suffered by the owners or users of property which had been contaminated as a result of a spill (consequential loss). One example of consequential loss is a fisherman's loss of income as a result of his nets becoming polluted.

An important group of claims comprises those relating to *pure economic loss*, i.e. loss of earnings sustained by persons whose property has not been polluted. A fisherman whose boat and nets have not been contaminated may be prevented from fishing because the area of the sea where he normally fishes is polluted and he or she cannot fish elsewhere. Similarly, an hotelier or restaurateur whose premises are close to a contaminated public beach may suffer loss of profit due to a reduced number of guests during the period of pollution.

Claims for pure economic loss are admissible only if they are for loss or damage caused by contamination. The starting point is the pollution, not the incident itself.

In order to qualify for compensation the basic criterion is that a sufficiently close link of causation exists between the contamination and the loss or damage sustained by the claimant. A claim is not admissible on the sole criterion that the loss or damage would not have occurred but for the oil spill in question. When considering whether the criterion of a sufficiently close link of causation is fulfilled, the following elements are taken into account:

- the geographic proximity between the claimant's activity and the contamination;
- the degree to which a claimant is economically dependent on an affected resource;
- the extent to which a claimant has alternative sources of supply or business opportunities and
- the extent to which a claimant's business forms an integral part of the economic activity within the area affected by the spill.

Account is also taken of the extent to which a claimant can mitigate the loss.

3. Environmental damage

In the 1992 Conventions "pollution damage" is defined as damage caused by contamination. The definition contains a proviso to the effect that compensation for impairment of the environment (other than loss of profit from such impairment) should be limited to costs of reasonable measures of reinstatement actually undertaken or to be undertaken.

Damage to the marine environment cannot be easily assessed in monetary terms, as the marine environment does not have a direct market value. In recent years models have been elaborated in many countries for the assessment of damage to the marine environment. It is submitted that any assessment of ecological damage to the marine environment in monetary terms would require sweeping assumptions regarding relationships between different components of the environment and economic values. Any calculation of the damage suffered in monetary terms would by necessity be arbitrary. For this reason, the 1992 Fund has taken the position that it would be inappropriate to admit claims for compensating damage to unexploited natural resources which have no owner.

The Funds have decided that in order for claims for the cost of measures to reinstate the marine environment to be admissible for compensation, the measures should fulfil the following criteria:

- the measures should be likely to accelerate significantly the natural process of recovery
- the measures should seek to prevent further damage as a result of the incident
- the measures should not, as far as possible, result in the degradation of other habitats or in adverse consequences for other natural or economic resources
- the measures should be technically feasible
- the costs of the measures should not be out of proportion to the extent and duration of the damage and the benefits likely to be achieved.

The assessment should be made on the basis of the information available when the specific reinstatement measures are to be undertaken.

Compensation is paid only for reasonable measures of reinstatement actually undertaken or to be undertaken, and if the claimant has sustained an economic loss that can be quantified in monetary terms. The Fund will not entertain claims for environmental damage based on an abstract quantification calculated in accordance with theoretical models. It will also not pay damages of a punitive nature on the basis of the degree of fault of the wrongdoer.

Studies are sometimes required to establish the precise nature and extent of environmental damage caused by an oil spill and to determine whether or not reinstatement measures are necessary and feasible. Such studies will not be necessary after all spills and will normally be most appropriate in the case of major incidents where there is evidence of significant environmental damage.

The Fund may contribute to the cost of such studies provided that they concern damage falling within the definition of "pollution damage" in the Conventions, including reasonable measures to reinstate a damaged environment. In order to be admissible for compensation it is essential that the post-spill studies are likely to provide reliable and usable information. For this reason the studies must be carried out with professionalism, scientific rigour, objectivity and balance. This is most likely to be achieved if a committee or other mechanism is established within the affected Member State to design and co-ordinate any such studies, as well as reinstatement measures.

The scale of the studies should be in proportion to the extent of the contamination and the predictable effects. On the other hand, the mere fact that a post-spill study demonstrates that no significant long-term environmental damage has occurred or that no reinstatement measures are necessary, does not by itself exclude compensation for the costs of the study.

The Fund should be invited at an early stage to participate in the determination of whether or not a particular incident should be subject to a post-spill environmental study. If it is agreed that such a study is justified the Fund should then be given the opportunity to become involved in the planning and in establishing the terms of reference for the study. In this context the Fund can play an important role in helping to ensure any post-spill environmental study does not unnecessarily repeat what has been done elsewhere. The Fund can also assist in ensuring that appropriate techniques and experts are employed. It is essential that progress with the studies is monitored, and that the results are clearly and impartially documented. This is not only important for the particular incident but also for the compilation of relevant data by the Fund for future cases.

It is also important to emphasise that participation of the Fund in the planning of environmental studies does not necessarily mean that any measures of reinstatement later proposed or undertaken will be considered admissible.

IV. Uniform application of the Conventions

The 1971 and 1992 Fund Assemblies have expressed the opinion that a uniform interpretation of the definition of "pollution damage" is essential for the functioning of the compensation regime established by the Conventions. In this regard the IOPC Funds' position applies not only to questions of principle relating to the admissibility of claims but also to the assessment of the actual loss or damage where the claims do not give rise to any question of principle.

The importance of uniformity of application is obvious. It is important from the point of view of equity that claimants are treated in the same manner independent of the State where the damage was sustained. In addition, the oil industry in one Member State pays for the cost of clean-up operations incurred and economic losses suffered in other Member States. Unless a reasonably high degree of uniformity and consistency is achieved, there is a risk of great tensions arising between Member States and of the international compensation systems no longer being able to function properly.

It should be noted that the definition of "pollution damage" is the same in the 1992 Civil Liability Convention and the 1992 Fund Convention. For this reason, the concept "pollution damage" should be interpreted in the same way independent of whether the claim is against the shipowner/the shipowner's insurer under the 1992 Civil Liability Convention or against the shipowner/his insurer and the 1992 Fund under both 1992 Conventions. Similarly, the concept should also be interpreted in the same way by the national courts whether the claim under consideration is under only the 1992 Civil Liability Convention or under both 1992 Conventions.

In May 2003 the 1992 Fund Administrative Council (on behalf of the Assembly) adopted a Resolution on the interpretation and application of the 1992 Civil Liability and Fund Conventions. In the Resolution attention was drawn to the importance for the proper and equitable functioning of the regime established by the 1992 Conventions that these Conventions were applied uniformly in all States Parties and that claimants for oil pollution damage were given equal treatment as regards compensation in all States Parties. The Resolution emphasised the importance that national courts in States Parties gave due consideration to the decisions by the governing bodies of the Funds on such matters.

V. Review of the adequacy of the international compensation regime

1. Increase in the limitation amounts available under the 1992 Conventions

When the 1992 Civil Liability and Fund Conventions were adopted, it was expected that the total amount available under these Conventions, at that time US $205 million, would be sufficient to compensate all victims in full, even in the most serious incidents. However, it became evident already in relation to the first major incident which occurred after the entry into force of the 1992 Conventions, namely the *Nakhodka* incident in Japan in 1997, that this was not the case. The inadequacy of that amount was demonstrated even more clearly in respect of the *Erika* incident in France in 1999.

In the light of this experience, a number of States took the view that it was necessary to increase significantly the amount of compensation available. A first step to this effect was taken in 2000 when the Legal Committee of IMO decided under a special procedure provided for in the Conventions (the "tacit amendment" procedure), to increase the limits contained in 1992 Civil Liability Convention and the 1992 Fund Convention by some 50%. The amendment to the 1992 Fund Convention brought the total amount available under the 1992 Conventions to US $310 million. The increases entered into force on 1 November 2003.

2. 1992 Fund Working Group

Many States took the view, however, that the increase in the maximum compensation amount decided by the IMO Legal Committee was insufficient and the point was made that although the system had worked well in most cases, there were inadequacies in the system and it was therefore necessary to carry out a general revision of the 1992 Conventions. For this reason the 1992 Fund Assembly established in 2000 a Working Group open to all Member States to examine the adequacy of the international compensation regime established by these Conventions.

The Working Group has held eight meetings, the most recent from 17 to 22 March 2005.

3. Supplementary Fund

During the Working Group's discussions it was decided to work towards the creation of an optional third tier of compensation and to prepare a draft Protocol providing for such a third tier by means of a Supplementary Fund. A Diplomatic Conference held under the auspices of the IMO in London in May 2003 adopted, after difficult negotiations, a Protocol creating such a Supplementary Compensation Fund. The Protocol entered into force on 3 March 2005. The main elements of the Protocol are as follows:

- The Protocol established a new intergovernmental organisation, the International Oil Pollution Compensation Supplementary Fund, 2003.
- Any State which is Party to the 1992 Fund Convention may become Party to the Protocol and thereby become a Member of the Supplementary Fund.
- The Protocol applies to pollution damage in the territory, including the territorial sea, of a State which is a Party to the Protocol and in the exclusive economic zone (EEZ) or equivalent area of such a State.
- The total amount of compensation payable for any one incident is 750 million SDR (US$1 140 million), including the amount payable under the 1992 Civil Liability and Fund Conventions, 203 million SDR (US$310 million)).
- Annual contributions to the Supplementary Fund are to be made in respect of each Member State by any person who, in any calendar year, has received total quantities of oil exceeding 150 000 tonnes after sea transport in ports and terminal installations in that State. However, the contribution system for the Supplementary Fund differs from that of the 1992 Fund in that at least 1 million tonnes of contributing oil will be deemed to have been received each year in each Member State for the purpose of paying contributions.
- The Supplementary Fund only pays compensation for incidents which occur after the Protocol has entered into force for the affected State.

Difficulties have arisen in some incidents involving the Funds where the total amount of the claims arising from a given incident exceeded the total amount available for compensation or where there was a risk that this might occur. Under the Fund Conventions, the Funds are obliged to ensure that all claimants are given equal treatment. In a number of cases the 1971 and 1992 Funds therefore have had to limit (pro-rate) payments to victims to a percentage of the agreed amount of their claims. In most cases it eventually became possible to increase the level of payments to 100% once it was established that the total amount of admissible claims would not exceed the amount available for compensation, but in many cases the delay in payment of part of the compensation nevertheless caused financial hardship to victims, for example fishermen and small businesses in the tourism sector. The 2003 Protocol will greatly improve the situation for victims in

States becoming parties to it. In view of the very high amount available for compensation of pollution damage in these States, it should be possible in practically all cases to pay all established claims in full from the outset.

4. Sharing of the financial burden between shipowners and the oil industry

When the Working Group discussed whether amendments should be made to the provisions in the 1992 Civil Liability Convention regarding shipowners' liability and related issues, it became clear that there was a great divergence of opinion.

The oil industry maintained that the international compensation regime should ensure that persons suffering oil pollution damage were compensated promptly but also be consistent with the general objective to improve maritime safety and reduce the number of oil spills. The point was made that the Supplementary Fund financed permanently by oil receivers would distort the balance between the shipowners' and oil receivers' contributions to the regime since it was financed only by oil receivers.

The shipowners and their insurers took the view that that the issues relating to shipowners' liability should not be reopened since to do so would be detrimental to the position of victims of oil pollution. It was suggested that the 1992 Conventions were intended to create an efficient compensation regime and had not been intended to ensure the quality of shipping or to punish the guilty party. It was emphasised that it was of paramount importance to maintain the equitable balance between the burdens imposed on the two industries involved, i.e. those of the shipping and cargo interests. The shipowners argued that the voluntary increase of the limitation amount applicable to small ships referred to below would preserve that balance.

Several options were put forward relating to the equitable sharing of the financial burden resulting from oil spills between shipowners and the oil industry but none of them received strong support.

Some delegations stated that none of the proposals for revision would result in more compensation becoming available to victims of pollution damage, that the only issues that needed to be resolved were the sharing of the financial burden between the shipping industry and oil cargo interests and the problem of substandard shipping and that both these issues were best addressed through industry initiatives. A number of delegations expressed the view that there was no justification for a further increase in the financial burden on shipowners beyond the 50% increase that came into effect on 1 November 2003 and the voluntary increase of the limitation amount for small ships set out below.

Other delegations expressed the view that although the Conventions had worked well in the past, there were serious deficiencies that went beyond financial considerations and the sharing of the financial burden, such as the need to ensure a quorum at meetings of the Fund's governing bodies, an effective means of enforcing oil reporting requirements and the uniform application of the Conventions. Those delegations therefore considered that a revision of the system was necessary and urgent. The point was also made that liability and compensation for oil pollution damage were matters of important policy considerations, which had to be governed by legislation.

5. STOPIA

The International Group of P&I Clubs[3] have, on a voluntary basis, increased the limitation amount for small tankers by means of an agreement known as the Small Tankers Oil Pollution Indemnification Agreement (STOPIA). The agreement came into force on 3 March 2005, ie the date of the entry into force of the Supplementary Fund Protocol.

STOPIA is a contract between owners of small tankers and applies to pollution damage in a State for which the Supplementary Fund Protocol is in force. It applies to all ships insured by one of the P&I Clubs that are members of the International Group and reinsured through the Group's pooling arrangement. Under the Agreement the maximum amount of compensation payable by owners of all ships of 29 548 gross tonnage or less would be US$30 million. Although the 1992 Fund is not a party to STOPIA, STOPIA confers legally enforceable rights on the 1992 Fund of indemnification from the shipowner involved in an incident.

The 1992 Fund will, in respect of ships covered by STOPIA, continue to be liable to compensate claimants if and to the extent that the total amount of admissible claims exceeds the limitation amount applicable to the ship in question under the 1992 Civil Liability Convention. The 1992 Fund would be entitled to indemnification by the shipowner of the difference between the shipowner's liability under that Convention and US$30 million, even if the Supplementary Fund is not called upon to pay compensation in respect of the incident.

6. Substandard transportation of oil

The Working Group has considered several proposals for dealing with the substandard transportation of oil. The intention of these proposals is to provide disincentives to shipowners to use substandard ships by imposing higher limits of liability on such ships. Under one proposal, cargo owners would also be liabile for pollution damage caused by such ships. Another proposal would deprive shipowners of their right to limit their liability if the incident resulted from structural defects of the ships (i.e. defects due to decay or lack of maintenance). No decision was taken on any of these proposals. Some States considered however that the issue of substandard shipping was not within the field of competence of the 1992 Fund but fell within the exclusive competence of the IMO and should be dealt with in the relevant IMO Conventions (SOLAS and MARPOL).

---

[3] Shipowners are normally insured for third party liabilities (including liability for oil pollution damage) in Protection and Indemnity Associations, so called P&I Clubs, which are mutual insurance associations owned by shipowners who are their members. Thirteen of these Clubs, which together insure some 95% of the world oil tanker fleet, form the International Group of P&I Clubs.

7. The Working Group's meeting in March 2005

At its meeting in March 2005 the Working Group focused its discussions on the equitable sharing of the compensation burden between shipowners and cargo interests and whether there was a need for a revision of the 1992 Conventions. The Working Group was evenly split as to whether the Conventions should be revised.

The Working Group also considered which issues to recommend to the Assembly for inclusion, if a limited revision of the 1992 Conventions were to be decided upon. It recommended that a number of issues should be retained for further consideration in the event the revision was to go ahead, and that others should be dropped.

The Assembly will consider the Working Group's report in October 2005 and decide whether or not the Conventions should be revised and, if so, which issues should be included in a revision.

VI. Concluding Remarks

The international compensation regime established under the Civil Liability and Fund Conventions is one of the most successful compensation schemes in existence. Most compensation claims have been settled amicably as a result of negotiations.

When the 1971 Fund was set up in 1978 it had only 14 Member States. Over the years the number of 1992 Fund Member States has increased to 93. It is expected that a number of States will ratify the 1992 Protocols in the near future. It is interesting to note that many States which have ratified the 1992 Conventions in the last few years were not previously parties to the 1969 and 1971 Conventions. This increase in the number of Member States appears to indicate that the Governments have generally considered the international compensation regime to be working well. This explains why the regime based on the 1992 Conventions has served as a model for the creation of liability and compensation systems in other fields, such as the carriage of hazardous and noxious substances by sea.

Although the Conventions were revised in 1992, the main features of the regime were decided in the late sixties and early seventies. It is not surprising therefore that the Contracting States have found that the regime needs to be revisited for modifications in the light of experience, so as to enable the regime to adapt to the changing needs of society and to ensure the regime's survival by remaining attractive to States. As a result of recent major incidents, the compensation regime based on the 1992 Conventions became subject to criticism for not providing adequate protection to victims of oil pollution, but the Fund's Member States have listened to this criticism and have taken it into account in a constructive way in the review of the adequacy of the regime which began in 2000.

Steps to improve the protection of victims of oil pollution have been taken resulting in the increases in the limits of liability and compensation which entered into force on 1 November 2003, the adoption in May 2003 of the Protocol establishing a Supplementary Fund and amendments to the Claims Manual in respect of the cost of post-spill studies and the costs of reinstatement of the polluted environment.

Annex

States Parties to both the 1992 Civil Liability Convention and the 1992 Fund Convention (as at 15 April 2005)

86 States for which the 1992 Fund Convention is in force (and which therefore are Members of the 1992 Fund)		
Algeria	Gabon	Oman
Angola	Georgia	Panama
Antigua and Barbuda	Germany	Papua New Guinea
Argentina	Ghana	Philippines
Australia	Greece	Poland
Bahamas	Grenada	Portugal
Bahrain	Guinea	Qatar
Barbados	Iceland	Republic of Korea
Belgium	India	Russian Federation
Belize	Ireland	Saint Vincent and the Grenadines
Brunei Darussalam	Italy	Samoa
Cambodia	Jamaica	Seychelles
Cameroon	Japan	Sierra Leone
Canada	Kenya	Singapore
Cape Verde	Latvia	Slovenia
China (Hong Kong Special Administrative Region)	Liberia	Spain
Colombia	Lithuania	Sri Lanka
Comoros	Madagascar	Sweden
Congo	Malta	Tanzania
Croatia	Marshall Islands	Tonga
Cyprus	Mauritius	Trinidad and Tobago
Denmark	Mexico	Tunisia
Djibouti	Monaco	Turkey
Dominica	Morocco	United Arab Emirates
Dominican Republic	Mozambique	United Kingdom
Fiji	Namibia	Uruguay
Finland	Netherlands	Vanuatu
France	New Zealand	Venezuela
	Nigeria	
	Norway	

7 States which have deposited instruments of accession, but for which the 1992 Fund Convention does not enter into force until date indicated	
Saint Lucia	20 May 2005
Malaysia	9 June 2005
Tuvalu	30 June 2005
Estonia	6 August 2005
South Africa	1 October 2005
Israel	21 October 2005
Saint Kitts and Nevis	2 March 2006

States Parties to the 1992 Civil Liability Convention but not to the 1992 Fund Convention (as at 15 April 2005)
(and therefore not Members of the 1992 Fund)

9 States for which the 1992 Civil Liability Convention is in force		
Bulgaria	Egypt	Romania
Chile	El Salvador	Switzerland
China	Indonesia	Vietnam

6 States which have deposited instruments of accession, but for which the 1992 Civil Liability Convention does not enter into force until date indicated	
Kuwait	16 April 2005
Solomon Islands	30 June 2005
Azerbaijan	16 July 2005
Saint Kitts and Nevis	7 October 2005
Syrian Arab Republic	22 February 2006
Pakistan	2 March 2006

States Parties to the Supplementary Fund Protocol (as at 15 April 2005)

8 States Parties for which the 2003 Supplementary Fund Protocol is in force (and which therefore are Members of the Supplementary Fund)		
Denmark	Germany	Norway
Finland	Ireland	Spain
France	Japan	

1 State which has deposited an instrument of accession but for which the Protocol does not enter into force until the date indicated	
Portugal	15 May 2005

Compensation by the Coastal States – The *Prestige* Disaster

Juan L. Pulido

I. Introduction

This paper is not intended to constitute an exhaustive treatment of all subjects related to the *Prestige* accident. Moreover, its purpose is not to provide an in-depth legal analysis of all relevant Spanish regulations ensuing from the *Prestige* accident, but simply to track the general trends of the legal policy. The paper will also attempt to elicit some of the lessons learned from the disaster.

With this perspective in mind, the paper will begin with a retrospective view of the accident and the damages caused. An analysis of the Spanish legislative measures adopted to cope with the spill will follow. Finally, I will focus on the general trends of that legislation.

II. The *Prestige* Disaster

On November 13, 2002, the single-hulled oil tanker *Prestige* was on passage in heavy seas and high winds in the Region of Cape Finisterre charged with 76.972 tonnes of heavy fuel oil. Only 25 miles off the Spanish coast, it began leaking oil.

The *Prestige* was refused safe harbour by Spain and Portugal, where she could have received shelter from the bad weather conditions, and where her cargo could have been pumped off.[1] Instead, she was ordered to go out to sea. The rough autumn seas of the Atlantic caused the ship to split in two and sink five days later, while being towed away from the coast, 130 miles off the coast of Galicia.

---

1 See IMO Resolution A. 949(23), December 2003, "Guidelines on Places of Refuge for Ships in Need of Assistance", Par. 1.3: "When a ship has suffered an incident, the best way of preventing damage or pollution from its progressive deterioration would be to lighten its cargo and bunkers; and to repair the damage. Such an operation is best carried out in a place of refuge." See also *Timagenis*, Places of refuge as a legislative problem: CMI Yearbook 2003 (2004) 375-379.

The leakage, break-up, and sinking of the vessel released an estimated 25 000 tonnes of cargo. Over the next weeks, oil continued to leak from the wreck at a declining rate.

The media impact of the *Prestige* affair was unprecedented, at least in Spain. Once more, a maritime disaster focused global attention on the problem of major oil tanker spills. The public reaction in Spain was tremendous: there was social shock, a number of popular movements seeking for compensation and punishment arose, and the Popular Party Government was ousted in the subsequent elections. The animosity held by those affected by the pollution was magnified by the way in which Spanish Authorities initially handled the accident.

The Spanish Parliament conducted an investigation into the disaster that reached no clear conclusion. Another political investigation is still in course in the Galician Regional Parliament.

In Galicia, as in Brittany, accidents involving oil spills are not a new occurrence. The population in these areas are regularly exposed to maritime spills: the *Polycommander* (1970), *Erkowitz* (1970), *Urquiola* (1976), *Andros Patria* (1978), *Casson* (1987), *Aegean Sea* (1992), etc.

III. Victims and Losses

The *Prestige* accident caused an environmental disaster of enormous scale. Due to the highly persistent nature of the fuel M-100, the released oil from the wreck drifted for extended periods with winds and currents, travelling great distances. An extensive area of the North Atlantic Spanish and French coasts were badly contaminated as a result of the fuel leak, more than 3 000 kilometres were polluted by the *Prestige* cargo. The spill contaminated one of the most productive ocean fisheries and shellfish beds in Europe, in a region whose main income is from fishing.

Damages resulting from the *Prestige* disaster are still being assessed. However, there are provisional figures that provide a precise picture of the scope of the catastrophe.[2] Information has been delivered by the 1992 Fund stating that, in May 2004, the Spanish Government estimated the total damage in Spain to be 834.8 million €; the French Government estimated the total damage in France to be 176 million €; and the Portuguese Government estimated the total damage in Portugal to be 3.3 million €. The Fund has determined the potential total claims exposure to be some 1038 million €.

By 22 February 2005, the Claims Handling Office in La Coruña had received 716 claims totalling 698 million €.

---

2 *Gonzales Laxe*, Análisis de las consecuencias económicas y sociales de los desastres marítimos: el caso del Prestige (2004).

Category of Claim	No. of Claims	Amount claimed €
Property damage	226	2441473
Clean-up	16	4161279
Mari culture	12	8026408
Fishing-shellfish gathering	147	134288947
Tourism	10	612472
Fish processors/vendors	257	12394672
Miscellaneous	44	1402843
Spanish Government	4	534695110
Total	**716**	**698023204**

IV. Ways for Compensation

1. Court Actions

There are pending inquiries, including criminal proceedings, against the captain of the vessel and Spanish maritime authorities.

a) Criminal Prosecution in Corcubión (Spain)

The Examining Court (Juzgado de Instrucción) of Corcubión is in course the preliminary proceedings (Diligencias Previas) 960/2002. The persons being criminally prosecuted are the master, the chief engineer and the first officer of the ship, together with the former General Director of the Spanish Merchant Marine, the Director of Operations of *Universe Maritime*, and the owner of the vessel. There are few doubts that the Spanish courts are competent.[3]

The captain is charged with ecological damage, and disobedience to Authorities (art. 325 y 326 Cod.Penal). He has been in prison several months. The judge's decision to lock up the captain is very controversial, since art. 230 UNCLOS bans the penalty of prison for pollution caused by foreign vessels. The only exception to this rule applies in cases of a wrongful act of pollution causing severe contamination of the territorial sea. The captain was finally released on bail for 3 million €, paid by the P&I Club (*The London Protection and Indemnity Club).*

The criminal prosecution of individuals can be criticized,[4] but it has to be understood in the context of the enormous, media-driven desire to blame individuals, and to urge the criminal prosecution of persons involved in the accident. Usually, and in this case, the captain is to blame.

3 See *Manjón-Cabeza Olmeda*, El caso Prestige – perspectiva jurídico-penal: Revista General de Legislación y Jurisprudencia, III época, 2002, no. 4, oct-dic., p. 573-593.

4 See, for example, *Simon*, La penalisation du droit est-elle efficace en matière de pollution marine?: Le Droit Maritime Française 645, February 2004, p. 166-168; *Marques*, La repression des rejets illicites d'hydrocarbures: Le Droit Maritime Française 647, April 2004, p. 307-323.

There is a lasting feeling, at least in public opinion, that the compensation system is not working, thus it is often the judiciary who fill the gaps using the Criminal Law.[5]

A further reason for this willingness to use the criminal justice system is to avoid the limits placed on liability. If someone is condemned in criminal proceedings, unlimited compensation results, as damages can be paid to anyone who can prove them. This is often the reason why mere accidents resulting from human error give rise to criminal charges.

Employing the criminal justice system in oil pollution incidents is not a new approach; it has been used in other jurisdictions, such as the USA, and also previously in Spain. In the *Aegean Sea* case, a Spanish criminal Court held that both the master and the pilot were criminally negligent and, in the same proceedings, found them jointly liable for the civil consequences of the grounding. As a result, the ship-owner, the Spanish state, The P&I Club, and the IOPCF were held directly liable. The same outcome resulted in France with respect to the *Erika* disaster.

b) Civil Liability Proceeding before the New York District Court

Brought by the Kingdom of Spain against ABS. The discovery of evidence is continuing. On 3 August 2004, the judge dismissed the counterclaim lodged by ABS against the Kingdom of Spain.

2. Claims against the 1992 Fund

a) The Fund Compensation in Previous Accidents[6]

Since it was set up, the IOPF has dealt with some 135 incidents in 25 countries. In most of the cases, the claims have been handled in a non-controversial way. The Fund has paid 710 million € of indemnities. For example, in the *Erika* case 79.4 million de € was paid in compensation, and in the *Prestige* incident 57.6 million was paid. Both cases are still in the process of being resolved. These accidents have clearly shown that the limits of indemnity are too low.[7]

b) Limits to Indemnity Applicable to the Prestige Accident

In the CLC92: limitation amount applicable 22 777 986 €. On 28 May 2003, the ship owner deposited the applicable amount in the Criminal Court of Corcubión for the purpose of constituting the limitation fund.

---

5 See Observations, Jurisprudence Françáise: Le Droit Maritime Française, 645, February 2001, p. 127.

6 See *Jacobsson*, Le régime international d'indemnisation des victimes des marées noires en pleine évolution: Le Droit Maritime Française 652, October 2004, p. 793-807.

7 Jacobsson (Note 6 supra) 797.

The maximum amount of compensation available by FUND92 in respect of the *Prestige* incident is 171 520 703 €, including the amount actually paid by the ship owner and his insurer.

c) Compensation in the Prestige Case

Until the final cost of the accident is known, the Funds can only pay a percentage of the eligible compensation claim (art. 5.4 Fund Convention 1992), in order to avoid overpayment or unequal treatment of claimants. Since the fund's limits have been exceeded on a number of occasions (e.g., the wreck of the *Nakhodka* in 1997, *the Erika* disaster in 1999), payments are made after long assessment procedures. The assessment procedures reduce the speed and amount of disbursements to victims.

At the 21st session held in May 2003, the Executive Committee of the 1992 Fund decided a compensation level of 15% for the *Prestige* oil spill, the lowest in the history of the 1971 and 1992 Funds.

Spain, France, and other affected countries argue that the approved compensation level leaves victims in an unsatisfactory situation. As a consequence, French and Spanish delegations have submitted a document to the Fund, to bring attention to the urgent need to improve payment levels. The Executive Committee of the Fund reviewed payment levels at the March 2005 session, but the Director felt unable to propose an increase in the level of payments beyond 15% for loss or damage suffered by the respective claimants.

V. Spanish Legislation in the Afternoon of the *Prestige*

In the aftermath of the *Prestige*, the problems that confronted the Spanish public authorities can be described as follows:

First of all, an accident with unprecedented economic consequences.

Secondly, a very delicate area of the coast and a poor region of the country particularly affected.

An inquisitive glare of the Media, because the handling of the accident by the competent authorities had been at least questionable.[8]

And finally, the evidence that the international current system for compensation would not pay the damages in full.

Since November 2002, but mainly in 2003, the *Prestige* oil spill caused a public outcry for legislation to prevent such accidents, and for rapid and equitable compensation for victims. The Spanish Government acted quickly (possibly too quickly), and passed several laws, statutes, and orders that formed a comprehen-

---

8 *Hetherington*, 'Prestige' – Can the Law Assist?: CMI Yearbook 2003 (2004) 361-374; *Shaw*, Places of Refuge. International Law in the Making: CMI Yearbook 2003 (2004) 329-343.

sive piece of legislation concerning many of the aspects of the spill, but mainly with compensation. These measures are still being passed.[9]

In the aftermath of the accident, it was completely out of question to wait to establish the final cost of the accident, and for compensation payments offered by the Funds. A decade before, the same region of Spain had suffered another tremendous maritime disaster, and the facts of that case remained present in everybody's mind. Regarding the *Aegean Sea* spill, claims established as admissible under the old regimen of 1971 had been paid out at 40 per cent.[10] In that case, the Spanish State criticised the handling of the claims on the basis that the payments were unreasonably low. A similar situation occurred with respect to the *Erika* disaster. In 2003 the system of the Fund had not yet permitted full compensation to the victims, when the accident occurred in 1999; the compensation was fixed at the beginning on a 50 per cent basis, later was increased, but it is still on a 80 per cent basis.[11]

Thus, it was widely known that the amount of compensation provided for in current international oil-pollution regimes would not satisfactorily address the extensive losses resulting from the *Prestige* incident. As a result, Spanish Authorities intervened by offering compensation to the victims, incurring considerable expenditures to combat the effects of the accident.

1. Laws

– Real Decreto-Ley 4/2004, 2 July, *por el que se adoptan determinadas medidas relacionadas con los daños ocasionados por el accidente del buque Prestige.*[12]

– Real Decreto-Ley 4/2003, 20 June, *por el que se introduce un sistema de anticipo de indemnizaciones al que podían acogerse voluntariamente los afectados por daños ocasionados en España como consecuencia del accidente del buque.*[13]

– Real Decreto-Ley 7/2002, de 22 de noviembre, *sobre medidas reparadoras en relación con el accidente del buque Prestige*[14]; endorse by the Congress of Deputies by Resolution of 28 November 2002.[15]

9 RD 276/2005, March 11 (BOE no. 61, March 12).

10 *Gaskel*, Pollution, Limitation and Carriage in the 'Aegean Sea', in Lex Mercatoria – Essays on International Commercial Law in honour of Francis Reynolds, ed. by Francis Rose (2000) 71, 77.

11 Reflexión de l'Académie de Marine sur la prévention des catastrophes maritimes: Le Droit Maritime Française 637, mai 2003, p. 454, 469; *Bulher*, Les marées noires: prévention et réparation : Le Droit Maritime Française 637, Mai 2003, p. 471, 474.

12 BOE no. 160, 3 July 2004.

13 BOE no. 148, 21 June.

14 BOE no. 281, 23 Novembre 2002.

15 BOE no. 290, 4 December 2002.

– Real Decreto-Ley 8/2002, 13 December, *por el que se amplían las medidas reparadoras en relación con el accidente del buque Prestige a las Comunidades del Principado de Asturias, Cantabria y País Vasco y se modifica el Real Decreto-Ley 7/2002, de 22 de noviembre;* endorsed by the Congress of Deputies by Resolution of 19 December.[16]

2. Regulations (selection)

– Real Decreto 1220/2002, 22 November, *por el que se crea la Comisión interministerial para el seguimiento de los daños ocasionados por el buque Prestige.*[17]

– Real Decreto 1/2003, 3 January , *por el que se crea el Comisionado para las actuaciones derivadas de la catástrofe del buque Prestige.*[18]

– Real Decreto 4/2003, 3 January, *por el que se modifica el Real Decreto 1220/2002, de 22-11-2002 (RCL 2002\2724), por el que se crea la Comisión Interministerial para el seguimiento de los daños ocasionados por el buque Prestige.*[19]

– Real Decreto 1053/2003, 1 August, *por el que se aprueba normas de desarrollo del Real Decreto-ley 4/2003, de 20-6-2003 (RCL 2003\1582), en relación con los daños ocasionados por el accidente del buque Prestige.*[20]

– Real Decreto 1341/2003, 31 October, *por el que se modifica el Real Decreto 1/2003, de 3-1-2003 (RCL 2003\40), por el que se crea el Comisionado para las actuaciones derivadas de la catástrofe del buque Prestige.*[21]

– Real Decreto 1699/2003, 12 December, *que Modifica el Real Decreto 1053/2003, de 1-8-2003 (RCL 2003\2005), por el que se aprueban normas de desarrollo del Real Decreto-ley 4/2003, de 20-6-2003 (RCL 2003\1582), en relación con los daños ocasionados por el accidente del buque Prestige.*[22]

– Real Decreto 102/2003, 24 January *que Dicta disposiciones complementarias para el funcionamiento de la Comisión interministerial para el seguimiento de los daños ocasionados por el buque Prestige.*[23]

---

16 BOE no. 310, 27 December 2002.
17 BOE no. 281, 23 November 2002.
18 BOE no. 4, 4 January 2003.
19 BOE no. 4, 4 January 2003.
20 BOE no. 184, 2 August 2003.
21 BOE no. 262, 1 November 2003.
22 BOE no. 298, 13 December 2003.
23 BOE no. 24, 28 January 2003.

– Real Decreto 276/2005, 11 March, por el que se desarrolla el artículo 2 del Real Decreto-ley 4/2004, 2 July, por el que se adoptan determinadas medidas relacionadas con los daños ocasionados por el accidente del buque Prestige.[24]

VI. Compensation Measures Adopted by the Spanish Authorities

Spanish Authorities have provided several forms of support to individuals and businesses affected by the spill, including payments, tax relief, and waivers of social security payments. The general principle underlying the legislative scheme is that the Spanish Government has made funds available to compensate in full all damages suffered by the victims of the spill.

1. Financial Compensation

Since the accident, the Spanish Central and Regional Governments have made several types of payments to victims.

It is important to distinguish between two different forms of financial compensation, which are different in nature and aim: grants, on the grounds of solidarity, and compensation, on the basis of the applicable law of damages.

a) Government Grants

Direct payments for economic loss. Disbursement on a daily basis to all those directly affected by the fishing bans, including shellfish harvesters, inshore fishermen and associated onshore workers with a high dependence on the closed fisheries, such as vendors, fish nets repairers and employees of fishing cooperatives, fish markets and ice factories.

40 €/day per fishermen and associated onshore workers. For ship owners, depending on the GT of their vessels, between 21 € per day, and 4.74 € per ton.

Reparation payments were initially intended to last for six months, but were extended until the end of the fishing bans, which lasted until October 8th 2003.

A common feature of these measures is that the requirements for being entitled to compensation are not very stringent; there are not precise requests about accreditation of damages, or specific indication about the kind of damages to be indemnified.

Under RDL 4/2004, the period in which members of the fishing industry could claim for direct losses suffered as a result of the accident was extended until 2004. However, the funds available for losses occurring in 2004 are limited to 3 million €. Claimants are required to submit their losses by 31 March 2005. Those entitled to compensation in 2004 are businessmen in the fishing, aquiculture, and shellfish harvesting sectors. To recieve compensation claimants must fill in a request with documentary evidence of the losses. If sufficient funds are available to pay all the

[24] BOE no. 61, 12 March 2005.

petitions, losses would be covered in full. However, if funding is inadequate, payments will be prorated.

The payments made by the government are not, strictly speaking, *compensation* in the sense of the CLC or FUND, but are financial aid, grants or subsidies provided by the state on the basis of solidarity. This distinction is expressly stated in the introduction of the RDL 7/2002 of 22 November, and in RD 276/2005. This principle of solidarity is included in the Spanish Constitution and it is a duty of public bodies.

b) Advanced Compensation Payments

In June 2003 the Spanish government adopted legislation making available 160 million € to compensate oil spill victims in full. To receive compensation, the claimants had to submit their claims by 31 December 2003. A precondition to the applicability of that measure was the Spanish State had to receive payment from the Fund.

In July 2004 the earlier regulation was modified to increase the amount available for compensation to 249.5 million €, and to extend the period for which compensation was available. The condition that the government was to receive payment from the Fund before full compensation to victims would be provided was repealed. The payments would come out of the public budget.

aa) Requirements to receive advance compensation

1. A *settlement agreement* must be signed with the Spanish State (art. 6). The agreement is voluntary. Victims who do not avail themselves of this proceeding would maintain their rights. Those individuals that choose to sign the settlement agreement accept the assessment of damages.
2. The damages alleged must be those covered by the definition of CLC92 and FUND92.
3. The damages have to be assessed (art. 4). The assessment of claims will be made according to the criteria applied in the CLC and FUND 1992.
4. The compensated victims have to renounce fully, unconditionally, and irrevocably the right to claim compensation in any other way in relation to the accident. Thus, for those receiving the funds, all judicial or extra judicial actions in Spain and abroad are ended.
5. The compensated victims also transfer the right of reparation to the Spanish Government. Once concluded, the agreement is a waiver of action by the victim and subrogation by the Spanish state of victim's rights.

bb) Financial support

The express aim of the measures is to provide an advanced payment with regard to the amount that has to be delivered by the 1992 Fund. Thereby, the Spanish Authorities have avoided a situation that would have obliged many of the victims to turn to the Funds for compensation.

After a general assessment of the total admissible damage in Spain carried out by its Director and the setting up of a financial guarantee by Spain, the 1992 Fund made a payment to the Spanish State of 57 550 000 €. All of this sum has been used to finance the system of advanced compensation payments for victims.

However a significant difference exists between the 249.5 million € provided by the Spanish State, and the 57.55 million € paid by the Fund. The Spanish Public Budget will cover this difference.

cc) Assessment of damages

The two damage assessment systems established under RDL 4/2003 and its developing rules are objective estimation and direct assessment.

Damage assessment by objective estimation is intended for those victims who received direct assistance when the accident occurred. Of these victims, more than 90% have already received compensation. This system has practically ended. It was aimed mainly at ship owners, crew members, shellfish harvesters, fish markets, etc. Approximately 86.2 million € has been paid to 22 800 claimants under this system.

The second system is direct assessment, which involves an individualised evaluation of the damage suffered. Direct assessment of damages is being conducted by the Consorcio de Compensación de Seguros (a Spanish State-owned insurance organisation), which works in close compensation with the Fund in order to apply the criteria of the 1992 Fund to the assessments. This system involved 5 000 claimants.

dd) People entitled to compensation

Every natural or artificial person suffering damage from the spill, which could be indemnified according to the rules of the CLC and FUND 1992, is entitled to compensation.

ee) The amount of the compensation

The Spanish regulation expressly states that its aim is to provide full compensation to the victims.

The first Decree stated that those entitled could receive "the corresponding part of the compensation to which they have right according to the payments received by the Spanish State," namely, that payments would be prorated.

In the second decree the requirement disappears, such that full compensation would be provided where the amount established for that is enough. Is the 249.5 million € sufficient? As a general rule, victims will not to receive full compensation from this source.

2. Loans

In art. 10 RDL 7/2002 the preferential lines of credit are laid down.

The *Instituto de Crédito Oficial* (ICO), which is a public body, has entered in credit operations totalling 100 million €, in anticipation of the reparation or reposition of the infrastructure of the fishing industry.

Payments would be in the form of direct loans and would have the following requirements:

1. Maximum loan amount – All damages must be justified by the delegation of the central government.
2. Entitlement – Natural and artificial persons qualify for loans, as well as small and medium enterprises as defined by the European Commission.
3. Term – The loan term is for one year, and can be extended.
4. Interest – The interest rate is 1.75%.
5. Guarantee – A guarantee can be established, but it is not compulsory.

3. Grants to Repair the Damages Caused by the Spill in Public Infrastructures

RDL 4/2003 states that the ICO will pay damages suffered by public bodies that enter into settlement agreements in order to repair public infrastructures. Spanish Government sources indicate 67 towns have requested compensation totalling 37.6 million €, and that the four autonomous regions affected have estimated their damages at 150 million €. The claimed amounts are awaiting approval by the State before payments are made to these bodies.

4. Waivers of Social Security Payments and other Labor Law Measures

a) Special Labour Law Measures

Art. 5.1 RDL 7/2002 establishes that in Redundancy Proceedings (*expedientes de regulación de empleo)* filed as a consequence of the *Prestige* disaster, the spill is to be considered an unavoidable circumstance, and can be a cause for the fair termination of a labour contract. The oil spill also constitutes grounds for a fair, temporary reduction of work hours.

In instances where a labour contract is terminated as a result of the oil spill, the *Fondo de Garantía Salarial*, a public body whose aim is to pay compensation in cases of employer bankruptcy, shall pay compensation.

b) Waivers of Social Security Payments

The RDL 7/2002 allows the *Tesorería General de la Seguridad Social* to grant employers a waiver of 100% of social security payments during the period of fishing bans. That period shall be considered as effectively paid. Self-employed workers that are included in the Special Regimen of the Social Security of the workers of the Sea (*Gents de Mer*) are also entitled to waive payment. Employers and self-employed workers that are entitled to waive social security payments, but have already made these payments, shall be reimbursed.

The sole requirement to receive this grant is the justification of the damages (without any other detail).

5. Tax Relief for Businesses Affected by the Spill

The aim of these measures is to reduce the impact of the spill on businesses that are directly or indirectly involved in the fishing, shellfish harvesting, and aquaculture industries.

a) Impuesto de Actividades Económicas (Local Tax on Businesses)

Art. 3 RDL 7/2002, November 23, reduces the Local Tax on industrial and commercial ventures, proportional to the period of inactivity caused by the fishing ban, for those artificial or natural persons *directly related* with the fishing industry. Municipal Bodies facing a loss of income as a consequence of this measure are to be compensated with funds from the State's General Budget.

A few weeks later, and surprisingly, a new Decree specified the reduction for the 2002 year period: a sixth of the tax for that year.

b) Income Tax

Art. 4 of RDL 7/2002 allowed taxpayers included in the objective assessment system of the Income Tax (IRPF), whose business or venture had been directly or indirectly affected by the fishing bans arising from the spill, to submit an application in request of a reduction of the applicable tax.

In December 2002, a new measure, RDL 8/2002, extended the benefits granted under RDL 7/2002 to include all taxpayers of Income Tax, regardless of whether they are included in the objective assessment system. The new measure is to be applied generally, without need of personal request, to victims of Galicia. However, taxpayers from other areas still need to submit a request.

All payments received as a result of the spill and established under RDL 8/2002 are tax-free, with respect to Income Tax calculation.

c) Value Added Tax (VAT)

RDL 7/2002 allows for businessmen *directly or indirectly* affected by fishing bans arising from the spill to file an application in request of a reduction of indexes for the tax assessment applied to their industry.

VII. General Principles of Law Underlying the Compensation Measures Adopted by the Spanish State

1. No Acknowledgement of Liability from the Part of the State

With respect to all of the compensation measures adopted by Spanish Public Bodies as a consequence of the *Prestige* oil spill, it is expressly stated that the payments do not constitute an acknowledgement of liability on the part of the paying body.

However, the Spanish state could incur liability resulting from the *Prestige* disaster. After the first leaking, the vessel is under the control of the Spanish coastal authorities. From then on, an erratic voyage began; the authorities ordered three different courses within six days. The actions taken by the Spanish authorities considerably increased in the risk to, and the damage suffered by the vessel, and extended the area affected by the spill. This conclusion has been established in a technical report from the *Universidad Politécnica de Barcelona,* a document that has been delivered to the competent judge of Corcubión.

Furthermore, the *Prestige* was refused safe harbour by Spain and Portugal, where she could have received shelter from the bad weather conditions, and had her cargo pumped off. Instead, the ship was towed out to sea. The decision to deny the *Prestige* safe harbour was, at least, questionable. The Spanish State may be liable to its own citizens and other States for acting negligently and failing to grant access to a place of refuge.[25] Moreover, the information provided by Spanish authorities was insufficient, and often wrong. Although the matter is still being resolved, there are grounds to conclude that Spanish Authorities may incur liability as a result of the accident.[26]

On what basis could the liability of the Spanish State arise?

a) In international maritime law, each state has the obligation to prevent harm to the marine environment of the exclusive zone of any other state and of the high seas beyond any of such zones (arts. 192, 194, 195, 221 and 235 UNCLOS).[27] This obligation translates into a duty to exercise the full measure of the state's territorial and extraterritorial legal authority to accomplish such prevention. As a result, each state must exercise their jurisdiction in pursuit of the state's general

---

25 *Ringbom,* You are welcome, but... Places of refuge and environmental liability and compensation, with particular reference to the EU: CMI yearbook 2004, 208, 216-218, pointed out that the obvious risk with accepting a ship to a place of refuge is that by directing it towards its own coastline, the State accepts a risk of pollution occurring in its waters, and it could result in contributory negligence on the basis of CLC article III(3). On the other hand, a refusal to accept a ship into refuge could involve legal or financial consequences for the coastal State. See also *Browne,* Places of Refuge- The IUMI Solution, IUMI Conference 2003, Seville, at <www.iumi.com/conferences/2003-sevilla/1609/BBrowne.pdf> p. 1-12; Report on Places of Refuge submitted by Comité Maritime International to the IMO Legal Committee: CMI Yearbook 2004, 389-393; *van Hooydonk,* Accommodating a ship in distress: rights and responsibilities of ports authorities: Towards a liability and compensation framework? Rights and responsibilities of port authorities, ESPO International Workshop, University of Antwerp, 11 December 2003 <www.espo.be/news/event_11-12-2003.asp>, p. 9-1.

26 *Vialard,* Faut-il reformer le regimen d'indemnisation des dommages de pollution par hydrocarbures?: Le Droit Maritime Française, 637, Mai 2003, p. 435, 447.

27 *Smith,* State responsibility and the marine environment (1998) 154; *Blanco-Bazan,* The Environmental UNCLOS and the Work of IMO in the Field of Prevention of Pollution from Vessels, in: International Maritime Environmental Law, ed. by A. Kirchner (2003) 31-47.

obligations. When a state fails to prevent its own representatives from causing environmental injury, legal responsibility may arise.

b) The *International Convention Relating to Intervention on the High Seas in Cases of Oil Pollution Casualties, of November 29 1969*, lays down the right of a coastal State to take such measures on the high seas as may be necessary to prevent, mitigate, or eliminate danger to its coastline from pollution by oil or the threat thereof, following upon a maritime accident.

In the *Prestige* case, the liability of the Spanish State will depend on the interpretation of art. 5 of the Convention. The coastal State is empowered to take only such actions that are necessary and reasonable, and after due consultations with appropriate interests including, the flag State or the States of the ship or ships involved, the owners of the ships or cargoes and, where circumstances permit, independent experts appointed for this purpose. Consultations should conclude as soon as possible. However, the construction of these rules is difficult, because of its vague wording. When is an action reasonable? The criteria included in the Convention are not sufficient to clearly answer this question. Some commentators take the position that the Spanish public bodies are clearly liable, mainly on the basis that the officials ordered the vessel's aberrant course.[28] This is likely the weak point of the Spanish position.

c) There are also grounds to support liability of the Spanish State under Spanish national law for wrongful functioning of public services: *Ley 30/1992* (RJAP), article 121 *Ley de Expropiación Forzosa, de 16 de diciembre de 1954,* and RD 428/1993.[29] As is the rule in *Civil Law countries*, Spanish law imposes liability on the State in the event of wrongful performance of the public duties. Spanish Law is based upon the general principle that a claimant may receive damages when a public body is at fault for causing the loss. The notion of *faute de service*, or linked to the public service, demarcates the circumstances in which the State will be liable for the actions of its servants. Similar rules operate in France[30], Italy, and Germany[31].

---

[28] *Beurier*, La sécurité maritime et la protection de l'environnement: évolutions et limites: Le Droit Maritime Française, 645, February 2004, p. 99, 110.

[29] See *Garcia Gomez de Mercado*, Legislación de Expropiación Forzosa, Comentarios y Jurisprudencia (2nd ed. 2001) 407-439. See, generally, *Muñoz Machado*, La responsabilidad concurrente de las Administraciones Públicas (1992); *Gonzales Perez*, La responsabilidad patrimonial de las Administraciones Públicas (1996); *Garrido Falla*, Tratado de Derecho Administrativo, Vol. II (1992).

[30] *Whittaker*, Principles of French Law (1998) 354; *Fairgrieve*, 16-19; *Grosdidier de Matons*, La responsabilité de l'Etat pour le fonctionnement de certains services maritimes: Le Droit Maritime Française 1968, p. 67-76; *Renard-Payen/Robineau*, La responsabilité de l'État pour faute du fait du fonctionnement défectueux du service public de la justice judiciaire et administrative, at <www. courdecassation. fr/_rapport/rapport02/etudes&doc/1-EtudeRenard-Payen.htm >.

[31] For Italy, see, *Duni*, Lo Stato e la responsabilità patrimoniale, (1968); Article 52 delle leggi sulla Corte dei Conti, passed on 12 June 1934, no. 1214; Article 18 dello Statuto degli impiegati civili dello Stato, Decree 10 January 1957. As regards Germany, see Section 839 of the Bürgerliches Gesetzbuch.

d) The Spanish State could also incur liability on the basis of art. 9.2 del FUND92, which states that if the Fund has paid a victim of pollution damage compensation for that damage, it will acquire by subrogation any rights the victim has against the person liable for the accident (art. 9).

e) Some Spanish scholars have even stated that grounds exist to find the Spanish government criminally liable with respect to negligence and crimes against the environment.[32]

There are many precedents of Judgments against States or Public Bodies, as a consequence of liability arising from a maritime accident. In Spain, the case of the *Urquiola* is a clear precedent for this principle. On May 12, 1976, while entering the port of La Coruña a tanker called the *Urquiola* struck an underwater rock and ran aground. The rocks did not appear in the official charts published by the Spanish state, despite the fact that two years earlier a master of another vessel had sent a communication indicating their existence to the competent authorities. The Spanish High Court, in a judgment of July 18, 1983,[33] ruled against the Spanish State, and awarded the claimants 10 million pesetas.

In summary, it may be that the Spanish State can be held liable for the *Prestige* oil spill, but responsibility for the accident would also lie with the flag State. Art. 217 of UNCLOS clearly indicates that liability for such accidents rests with the flag state. While flag states have an obligation to exercise jurisdiction over their vessels to prevent marine pollution, the standard of performance of this obligation remains undefined.[34] The French Academy of Marine Merchant has recently recommended that states affected by an oil spill should have recourse against the vessel's flag state.[35] Some measures consistent with this obligation have been proposed recently, e.g., the ban to enter in the Ports of the affected State for vessel of the flag state.[36]

Furthermore, is it not the time to put increased pressure on the IMO in order to promote an international set of flag states obligations?

2. Subrogation in the Rights of the Compensated Victims

Under Spanish statutory regulations, if a public authority has paid any compensation to an individual for pollution damage, it will acquire by subrogation any rights the recipient of the payment would have had against the Fund, or against third parties. The Decrees expressly state that compensation payments will be

32 See, for example, *Martinez-Bujan Perez*, La posible responsabilidad penal del gobierno: El País, Saturday, 1 March 2003, p. 26.

33 Sentencia Tribunal Supremo 18 July 1983, Aranzadi 4085.

34 *Smith*, State responsibility and the marine environment (1998) 155; *Daniel*, Civil Liability Regimes as a Complement to Multilateral Environmental Agreements: Sound International Policy or False Comfort?: (2003) 12 Review of European Community & International Environmental Law 225, 238; *Vialard* (Note 26 supra) 446.

35 Reflexión de l'Académie de Marine sur la prévention des catastrophes maritimes (Note 11 supra) 460.

36 Bulher (Note 11 supra) 475.

included in subrogated claims by the Spanish authorities, pursuant to article 9.3 of the Fund Convention 1992.

3. Full Compensation

Under the existing international regime, the maximum compensation limits set out are insufficient to provide full compensation to victims. For that reason the Spanish regulations, at least the first decrees passed immediately after the spill, underlined that the aim of the measures was to offer full compensation to the victims. However, the goal of fully compensating injured parties was soon derailed by new, aforementioned measures specifying the actual scope of the compensation measures.

Nevertheless, although it cannot be said that the Spanish State has provided full compensation for all damages, most of the victims have received adequate payment, and the payments are higher than the amount expected from the Funds. In short, the aim of the current Spanish compensation measures is to offer, to the extent possible, full compensation to victims.

As a consequence of the compensation measures, the expenses incurred by the Spanish State far exceed any limits set up by the current international legislation. That being so, the low level of compensation provided by the Fund directly affects the Spanish State. Furthermore, the Spanish government has incurred other expenses that are not strictly speaking compensation payments, but whose aim is to offer economical aid to the victims of the spill, e.g., tax exemptions, the direct aid, or investment in affected areas. It is likely that the Spanish State will pay damages arising from the accident, whether or not it receives compensation from the Fund, the ship owner, or even from the classification society. As a result, the Spanish taxpayer will ultimately pay a large share of the bill for the *Prestige* disaster.

4. All Damages

At the beginning, it was clear that the intention underlying the Spanish legislative measures was to cover all damages arising from the spill, dispensing with the concept of damage set out in current international oil pollution law. In the aftermath of the *Prestige* incident, Spanish public opinion left no room for subtleties. Claims appeared everywhere, and the Spanish public bodies had neither the strength nor the will to distinct or to investigate the degree of proximity of those casualties. That is why the first Spanish statutory regulations provided compensation for damages *directly and indirectly* related to the accident.

RDL 7/2002 and RDL 8/2002, contain articles expressly stating that the Commission, created to manage the accident, will assess the indirect damages resulting from the spill, and will propose to the Government the means to compensate them. Some of the financial measures, such as those providing tax relief, are aimed at compensating direct and indirect damages. It is also significant that the standard form Settlement Agreement, as between the victims and the Spanish Authorities, expressly states in the Preamble that: "Both parties accept as assessment of the damage directly or indirectly arising from the spill of the Prestige."[37] Finally, many of the payments and investments incurred by the Spanish public authorities are compensation for indirect damages or, at least, payments of damages that would not meet the requirements of the concept laid down in CLC and FUND92.

In summary, although there are some contradictions in the aforementioned legislative measures, it can generally be said that the Spanish regulations concerning the *Prestige* disaster tend to compensate a broader range of damages than those included in the definition of the CLC and Fund 1992.

VIII. Concluding Remarks

1. The Spanish regulations concerning the *Prestige* oil spill have broadened the compensation scheme set out in the current international civil liability regime for marine oil pollution to offer *almost* full compensation for *almost* all types of damages. This paper has tried to show that the Spanish public bodies have adopted some very special legal provisions and administrative mechanisms to deal with the *Prestige* problem.

The Spanish State, under the enormous pressure of public opinion, was compelled to facilitate prompt, full, and equitable compensation payments to victims suffering damage in their territory. The strength of the environmentalist sentiment in the aftermath of the *Prestige* incident has also led to the adoption of preventive measures, which are sometimes contrary to current international maritime law (e.g., the ban of single hull tankers).

Although the 1992 reforms to the CLC and FUND tend to broaden the concept of pollution damage to include compensation for loss of profits, the extended definition still does not give rise to full compensation for victims of an oil spill like that of the *Prestige.*

As a result, the Spanish State, acting as guardian of collective interest, has offered more generous compensation to victims of the oil spill. It can be reasonably expected that at least a significant percentage of the amount of compensation paid by the Spanish government to victims will be recovered from the Fund.

The Fund has recognised that public bodies can be legitimate claimants under the oil pollution liability regime, but the damage claims submitted must be related to quantifiable economic damage. As a result, many of payments made by the Spanish State would be irretrievable. And thereby, it should be stressed that the Spanish State has been the *Deep Pocket* of the *Prestige* disaster.

The Spanish State has fulfilled the implicit obligation laid down in article 235 of UNCLOS, requiring States to assure prompt and adequate compensation in respect of all the damages resulting from marine pollution. The same obligation is also set out in art. 45 of the Spanish Constitution, which states that Spanish public

[37] ORDEN HAC/114/2004, 27 January 2004.

bodies have the obligation of ecological rehabilitation, and must also prosecute trespassers.

The solutions and methods chosen by the Spanish authorities represent a very generous attempt to offer full compensation to all victims of the accident. However, while some of the payments made by the Spanish State are, strictly speaking, compensation for civil liability, others are grants or subsidies arising from the constitutional duty of solidarity.

The system the Spanish authorities employ is not perfect, but it is rooted in political and justice considerations, namely the need to offer full and prompt compensation for victims of pollution in a region frequently punished by such accidents.

2. In some ways, an alternative compensation mechanism has been employed in the *Prestige* case: the State pays victims and subrogates their rights, and then tries to recover from the Fund, the ship owner, or the insurer. Would it be a sensible idea to incorporate this solution into the legal regimes governing oil pollution compensation?

In fact, this type of compensation system, whereby the State compensates victims in full, and subsequently recovers from the Fund, has recently been proposed. This system would be more effective and equitable, and would be particularly efficient when the administrative liability of the State is questionable.[38]

This system also has shortcomings, such as the scheme relying on the government to voluntarily undertake to compensate victims, which in some cases would be unacceptable.[39]

Some commentators have proposed another compensation mechanism,[40] whereby an international Fund would pay full and immediate compensation to the victim, and the Fund would then have recourse against the persons responsible for the spill. This is an interesting idea that deserves further thought.

3. It is evident that the current environmental liability rules are insufficient to meet the compensation claims arising from oil spill incidents by the affected public. When confronted by the real risks of insufficient compensation, the Spanish Government decided to offer more extensive compensation.

This situation underlines some of the shortcomings of the existing international oil pollution regime with respect to the issue of responsibility. A good example of this relates to the problem of the limits of liability. The history of oil-pollution liability regimes shows that the conventions seem to be reactive, in the sense that they are successively amended after each new incident when it is a proven fact

[38] *Bulher* (Note 11 supra) 475; *AA.VV*, Problemas jurídico-administrativos planteados por el Prestige (2004) 97.

[39] European Commision's Explanatory Memorandum of the Proposal of the COPE fund; see in *Ringbom*, The 'Erika' accident and its effects on EU Maritime Regulation, in: Current Maritime Environmental Issues and the International Tribunal for the Law of the Sea, ed. by Nordquist/Moore (2001) 265, 277.

[40] *Vialard* (Note 26 supra) 435-451.

that the previously agreed limits are insufficient to meet greater oil spills. This is a recurring problem, as there are few doubts that new accidents will occur in the future, which will exceed the level of compensation laid down in the current regime. A consequence of this situation is that the burden of the risk of the dangerous activity is shared among those who are not polluters, in our case, the state.[41] Furthermore, it is not the polluters who pay, but taxpayers and consumers.

4. Recently, it has been questioned whether a special legal regime for large-scale accidents is needed. The main problem with the liability limits under the CLC and the FUND, in the 1969 and 1992 schemes, is that claims resulting from major accidents regularly exceed the limits set. This presents a real barrier to the early settlement of claims, as it is not clear at the outset whether the claims will be prorated.[42] The sinking of the *Prestige*, in particular, has cast doubt on whether the existing regime is indeed sufficient to deal with major oil spills.

5. Finally, it is evident that something must be changed in the current regime of oil pollution damages: not only the ancillary questions, but the core issues. Only if this demand is fulfilled, will recourse to unilateral measures cease. These measures are the response to an ever-increasing public demand, and are the method used to fill the gaps left by the current international law.

The supplementary fund and the increasing limits of liability are important improvements that would promote a better compensation of damage. However, the current system's core problems are still to be solved. In particular, the basis and limits of liability, the channelling of the responsibility, and the definition of damage.

6. The way in which Spanish Authorities handled the *Prestige* oil spill can be criticised in many ways. However, the Spanish State's response to the accident, with all its imperfections and hastiness, has to be understood: to cope with such a marine disaster is a tremendous challenge for any government. People living in the most exposed parts of the Spanish coasts no longer have any tolerance for oil spills, and there was massive pressure on the government to provide instant, unilateral solutions.

---

41 *Daniel* (Note 34 supra) 240.

42 *Gaskel* (Note 10 supra) 76.

International Convention on Civil Liability for Bunker Oil Pollution Damage, 2001 – Liability and Insurance Aspects

Ling Zhu

I. Introduction

The International Convention on Civil Liability for Bunker Oil Pollution Damage, 2001 (the Bunkers Convention) was adopted by the International Maritime Organization (IMO) in March 2001. The aim of the Bunkers Convention is to ensure the availability of adequate, prompt and effective compensation to persons who suffer pollution damage caused by oil spills when the oil is carried as fuel in ships' bunkers.[1]

Prior to the Bunkers Convention, the IMO has adopted several conventions regarding civil liability and compensation for oil pollution damage. These earlier conventions are the 1969 International Convention on Civil Liability for Oil Pollution Damage (1969 CLC), the 1971 International Convention on the Establishment of an International Fund for Compensation for Oil Pollution Damage (1971 Fund Convention), as well as their 1992 Protocols. In March 2005, the 2003 Protocol on the Establishment of a Supplementary Fund for Oil Pollution Damage also entered into force. However, these conventions do not deal with liability issues concerning oil spills from the vessels other than oil tankers.[2]

---

1 The Bunkers Convention states in its preamble: "Considering that complementary measures are necessary to ensure the payment of adequate, prompt and effective compensation for damage caused by pollution resulting from the escape or discharge of bunker oil from ships."

2 Basically, "ship" was defined in Article 1(1) of the 1969 CLC as "…any sea-going vessel and any seaborne craft of any type whatsoever, actually carrying oil in bulk as cargo." This definition was replaced in its 1992 Protocol with the following text: "'Ship' means any sea-going vessel and seaborne craft of any type whatsoever constructed or adapted for the carriage of oil in bulk as cargo, provided that a ship capable of carrying oil and other cargoes shall be regarded as a ship only when it is actually carrying oil in bulk as cargo and during any voyage following such carriage unless it is proved that it has no residues of such carriage of oil in bulk aboard." Apparently, the definition of "ship" confines the application scope of the 1969 CLC and its Protocol

Oil spills from ships other than oil tankers may cause expensive pollution damage to the marine environment.[3] For instance, of the 12 largest oil spills in Australia,[4] the government's response costs for oil tanker spills averaged US$115,000, while response costs for spills from non-tankers (generally heavy fuel oils) averaged over five times greater at US$625,000; three of the four most expensive spills in terms of response costs were those from heavy fuel oil from non-tankers.[5] The *Pallas* incident in Germany showed that tremendous damage might be caused by a limited amount of fuel oil leaking from a cargo vessel.[6] In addition, it is estimated that on average the amount of fuel in bunkers carried by non-tankers is around 14 million tons at any given time – compared with approximately 130 million tons of oil carried as cargo on the world's seas.[7] Some bulk carriers and container ships carry more oil as bunker fuel than tankers carry as cargo.[8] It can thus be reasonably assumed that if a very large container ship lost her entire bunker load, the resultant pollution damage would, in qualitative terms, equate to the loss of a fair sized tanker, subsequently involving a large compensation payment.[9]

On a national level, legislation dealing with bunker-oil pollution liability is in force in some countries. For instance, the United States adopted the Oil Pollution Act 1990, which deals with oil pollution from all types of vessels.[10] However,

---

within oil tankers. Under the CLC Protocol 1992, bunker-oil spill may be covered from such a ship when it is sailing in ballast whilst merely employed in the oil trade, not when it is engaged in carriage of other types of cargoes. It was considered that there is a clear difference between oil carried as cargo and as bunker fuel oil. Accordingly, the earlier conventions are not applicable to the spills of bunker oil from ships other than tankers.

3 The high response costs involved in a bunker-oil spill incident is mainly due to the nature of the bunker oil. The experience shows that many fuel oil used as ships' bunkers is the most difficult oil to combat, because they are highly viscous and persistent.

4 These oil spills likely occurred between 1975 and 1997, see IMO LEG75/5/1.

5 See IMO LEG 75/5/1.

6 The fuel oil was spilled from the Bahamian wood carrier *Pallas*, which drifted aground off Germany's Amrum Island in the North Sea in 1998. This oil spill affected nearly 30,000 sea birds, and environmental groups predicted that the spill could eventually impact more than 100,000 birds. For further information see: <http://www.waddensea-secretariat.org/news/publications/Wsnl/Wsnl99-1/articles/09-reineking.pdf>.

7 IMO Legal Committee: LEG 75/5/1.

8 The information is available at <http://www.imo.org/Newsroom/contents.asp?topic_id=67&doc_id=457>. See also *DelaRue/Anderson*, Shipping and the Environment (1998) 263: "...although only oil tankers can cause very large spills, oil tankers are not the only ships carrying pollutants. Many bulk carriers and container ships carry bunker fuel of 10,000 tons or more, and there are large quantities than many of the world's tankers carry as cargo".

9 The same opinion was mentioned in 'Leading Article: Good Oil', Lloyd's list, 14 October 1998.

10 The Oil Pollution Act 1990, Section 2701(37): "'vessel' means every description of watercraft or other artificial contrivance used, or capable of being used, as a means of transportation on water, other than a public vessel".

many countries do not have domestic legislation to deal with liability and compensation for bunker-oil pollution damage. If the legislation in this respect is not uniform at the international level, the involved parties may be unclear as to the extent of their rights and liabilities. For instance, shipowners would be exposed to divergent domestic laws. Thus, they might decide to flee to the jurisdiction with less demanding legislation. Another consequence might be that the pollution victims will face difficulties in finding the liable person and in choosing the jurisdiction to file the claim and so on.

As explained, the established liability conventions on oil pollution limit their application to oil tankers. The study in this respect shows the possible serious loss or damage which can be caused by a bunker-oil spill. As the result, the Bunkers Convention was adopted. The Bunkers Convention, which has not yet entered into force, is a freestanding instrument and is largely modelled on the earlier international civil liability conventions.

II. Liability and Insurance Aspects under the Convention

It remains to be seen whether the Bunkers Convention will benefit pollution victims of bunker-oil spills since it has not come into force. In order to see the overall picture of this Convention, four points are worth mentioning:

First, different types of ships are covered in the Bunkers Convention for the purpose of dealing with bunker-oil pollution liability; however, the Bunkers Convention does not apply to oil tankers.[11]

Second, the oil causing the pollution damage must be the "bunker oil" as defined in the Bunkers Convention. Bunker oil is defined as "any hydrocarbon mineral oil, including lubricating oil, used or intended to be used for the operation or propulsion of the ship, and any residues of such oil."[12]

Third, without much debate, the rule of strict liability is adopted so that the shipowner is liable for pollution damage irrespective of the existence of any fault on his side. The shipowner includes "…the registered owner, bareboat charterer, manager and operator of the ship."[13] The reason for imposing strict liability on the shipowner is two-fold: (1) strict liability has been adopted in other international civil liability conventions; (2) a different type of liability regime may have placed victims in a disadvantageous situation. Maritime transportation is an activity over which victims have no control. In oil spill incidents, the causal link between the incident and pollution damage may be very obscure. As a result, it is difficult for victims to identify the liable persons and to establish their claims. The strict liability rule facilitates the victims' claims to a great extent. The shipowner, however, may have the right of recourse independently of the Bunkers Convention. The right of the shipowner to recover from third parties is expressly preserved by the

[11] The Bunkers Convention, Article 1(1).
[12] The Bunkers Convention, Article 1(5).
[13] The Bunkers Convention, Article 1(3).

Bunkers Convention.[14] He can also benefit from immunity outside the Bunkers Convention.[15] In addition, when more than one person is liable in an incident, all liable persons falling within the definition of the "shipowner" shall be found jointly and severally liable.[16]

Fourth, the Bunkers Convention requires the registered owner of a ship to take out insurance or some other financial security to cover his liability for oil pollution damage.[17] However, neither "compulsory insurance" nor "other financial security" is defined in the Bunkers Convention. In terms of the relevant provisions, "compulsory insurance" in the Bunkers Convention not only requires the parties involved to purchase insurance policy and provide the evidence that they have done so, but also specify the insurance requirements relating to: (1) specified liabilities; (2) a certain amount and (3) the liability insurer, who can be sued directly by claimants.

Furthermore, a detailed analysis will be given to five issues relating to the liability and insurance for bunker-oil pollution damage. They are:

(1) No separate liability limitation regime for bunker-oil pollution liability;
(2) Not all liable parties are required to purchase insurance;
(3) The challenges for the P&I Clubs as the main liability insurer;
(4) Insurance and the question for adequate compensation; and
(5) Insurance certificate and its validity.

1. No separate limitation regime for bunker-oil pollution liability

A claimant-oriented liability system should not limit shipowners' liability to the extent that genuine claimants are deprived of sufficient compensation.[18] For this reason, a set of separate limitation rules respecting tanker-oil pollution liability is provided in earlier international civil liability conventions, which provides higher limit of liability than that would be available under general limitation conventions.[19] The Bunkers Convention confirms the shipowner's right to limit his liability; however, the extent of limitation relies on "any applicable national or international regime."[20] Apparently, this provision is not intended to establish a sepa-

[14] The Bunkers Convention, Art. 3(6): "Nothing in this Convention shall prejudice any right of recourse of the shipowner which exists independently of this Convention".

[15] The Bunkers Convention, Article 3(5): "No claim for compensation for pollution damage shall be made against the shipowner otherwise than in accordance with this Convention".

[16] The Bunkers Convention, Article 3(2): "Where more than one person is liable in accordance with paragraph 1, their liability shall be joint and several".

[17] The Bunkers Convention, Article 7(1).

[18] *Grime*, Implementation of the 1976 limitation convention – liability for maritime claims: (1998) Marine Policy, 306, 308.

[19] The general limitation conventions here refer to the Convention on Limitation of Liability for Maritime Claims, 1976 and its 1996 Protocol.

[20] The Bunkers Convention, Article 6: "Nothing in this Convention shall affect the right of the shipowner and the person or persons providing insurance or other financial

rate limitation regime or a limitation fund available to be exclusively devoted to bunker-fuel pollution claims.

A resolution respecting the limitation of liability was also affirmed upon the adoption of the Bunkers Convention. This resolution urges States that have not done so, to ratify, or accede to the Protocol of 1996 to amend the Convention on Limitation of Liability for Maritime Claims, 1976 (LLMC Protocol 1996), [21] which entered into force in May 2004. This resolution, to some extent, tries to make the applicable limitation rule for liability certain and uniform when implementing the Bunkers Convention. However, this resolution is only a recommendation; States are free not to accept the resolution and maintain their legislation in respect of the limitation of liability after the ratification of the Bunkers Convention. As a result, some countries may have failed to ratify any limitation convention, or have enacted different limitation regimes, or fail to put in place domestic legislation. It is even possible that the shipowner in a bunker-oil spill incident needs to undertake unlimited liability if the Contracting States to the Bunkers Convention requires it to do so.[22] Furthermore, even if all the States involved in a bunker-oil spill incident may be party to the LLMC Protocol 1996 or any earlier limitation conventions, problems may still arise from the absence of a specific limitation regime. For instance, it is unclear whether adequate funds to compensate all victims who have suffered pollution damage would be provided. Moreover, it is yet to be determined whether different pollution claims arising from a bunker-oil spill would be eligible for the limitation of liability under the LLMC.

2. Not all liable parties are required to purchase insurance

The Bunkers Convention holds the "registered owner, bareboat charterer, manager and operator of the ship"[23] liable for oil pollution damage.[24] However, the insurance requirement is imposed only on the registered owner of a ship,[25] and only if

---

security to limit liability under any applicable national or international regime, such as the Convention on Limitation of Liability for Maritime Claims, 1976, as amended."

21 There are two other resolutions which were adopted at the same time: (1) the Resolution on promotion of technical co-operation, and (2) the Resolution on protection for persons taking measures to prevent or minimize the effects of oil pollution, the details are available online at <http://www.imo.org/Conventions/mainframe.asp?topic_id=256 &doc_id=666>.

22 *Tsimplis*, The Bunker Pollution Convention 2001 – Completing and harmonizing the liability regime for oil pollution from ship?: (2005) LMCLQ 83, 83: "Thus, in States Parties where no general limitation of liability is available to the shipowner, the strict liability established by the Convention and the required compulsory insurance would, presumably, have to be unlimited".

23 The Bunkers Convention, Article 1(3).

24 The Bunkers Convention, Article 3(1).

25 There are other types of financial security institutions, which can satisfy the compulsory insurance requirement under the Convention. This aspect will not be discussed in detail here.

the ship has a gross tonnage greater than 1 000 and is registered in a State Party.[26] The provision leaves other parties falling within the definition of "shipowner" and ships with a gross tonnage of less than 1 000 outside the compulsory insurance requirement. However, the registered owner's liability insurance does not cover other parties who may be liable under the Bunkers Convention, and it remains a matter of choice whether such parties take out insurance.

3. The challenges for the P&I Clubs as the main liability insurer

Most of the shipowner's third party liabilities, including oil pollution liability, are insured by the shipowner's P&I Clubs. In the IMO legal committee meeting, the International Group of P&I Clubs agreed to continue providing insurance coverage to satisfy the compulsory insurance requirement under the Bunkers Convention.[27] However, since the Bunkers Convention covers many different types of ships, difficulties not experienced when providing insurance under earlier international civil liability conventions will likely be encountered under this new insurance arrangement.[28] Most evidently, to satisfy the insurance requirement under the earlier international civil liability conventions, the P&I Club must provide necessary evidence to show the insurance coverage for oil pollution liability is in place as required. This enables the relevant governmental authority to issue a certificate.[29] Accordingly, more paper work will be required for the P&I Club to assist the shipowner in getting his certificate issued under the Bunkers Convention.

The P&I Club's capacity to insure oil pollution liability is based on two factors: (1) reliance on the principle of mutuality, and (2) the existence of the International Group of P&I Clubs. "Mutuality" means sharing interests and risks with those in the Club. In terms of liability for oil pollution risk, mutuality refers to the notion that each member is liable to pay a share of the pollution costs resulting from

26 The Bunkers Convention, Article 7(1). In addition, Article 7(12) provides that, "Subject to the provisions of this Article, each State Party shall ensure, under its national law, that insurance or other security, to the extent specified in paragraph 1, is in force in respect of any ship having a gross tonnage greater than 1 000, *wherever registered*, entering or leaving a port in its territory, or arriving at or leaving an offshore facility in its territorial sea". Therefore, in effect, even if the ships are not registered in a State Party, they are required to maintain insurance or other financial security once falling with the Article 7(11).

27 See IMO LEG/CONF.12/9.

28 This is under the 1969 CLC and its 1992 Protocol.

29 *Rosag*, Compulsory marine insurance, originally published in: Scandinavian Institute of Maritime Law Yearbook 2000, available online at <http://folk.uio.no/erikro/WWW/corrgr/insurance/simply.pdf>: "...each vessel needs a paper certificate on board, which must be renewed regularly. Each renewal involves the P&I club, that issues a so-called blue card, the governments, that issue or authorize the certificate and scrutinize the insurer, and finally a logistics problem in getting the certiicate on board the vessel in time".

damage caused by himself, and all other members of the Club.[30] The system of levying calls, rather than charging premiums, is the central feature of the concept of mutuality.[31] The "call system" is comprised of the advance call, the supplementary call, and the overspill call in a catastrophic year. Members in the Club are obligated to pay the supplementary call to the extent that the advance call is insufficient to cover the claims. The operation of this system is regulated by the Club Rules. Furthermore, the P&I Clubs have developed an "international group" in which the Clubs share claims with each other, and purchase high levels of reinsurance on a collective basis.[32] The purpose of establishing an international group is to set up "claims-sharing agreement", whereby the claims made by one Club in excess of a certain amount are shared proportionately among all the Clubs in the pool.[33] Accordingly, it can reasonably be argued that the Club can potentially absorb large claims. However, its capacity is not indefinite and it routinely maintains a limitation for oil pollution liability. The Clubs in the International Group offer a limitation for oil pollution liability which currently stands at US $1 000 million for each accident or occurrence in respect of each ship entered by or on behalf of an owner not being a charterer other than a demise or bareboat charterer. This was in accordance with the decision of the International Group to increase reinsurance protection for oil pollution in 2000.[34] This limitation also applies to liability arising from any bunker-oil pollution damage. The limitation is not necessarily disadvantageous *per se*; however, considering that second-tier compensation does not exist,[35] and that the P&I Club is the liability insurer for most of shipowners, this limitation can be prejudicial to pollution victims receiving full compensation.

---

30 *Bennett*, Mutual risk – P&I insurance clubs and maritime safety and environmental performance: 25 (2005) *Marine Policy*, 13, 14.

31 *Hazelwood*, P&I Clubs Law and Practice (2000) 121, footnote omitted.

32 See the list of constituent Clubs at <http://www.ukpandi.com/ukpandi/infopool.nsf/HTML/About_IG>.

33 *Hazelwood* (Note 31 supra) 385.

34 The information is from <http://www.ukpandi.com/ukpandi/Infopool.nsf/HTML/E8EBE7157D236C5080256DB30055E4BD?Open&Highlight=2,oil%20pollution,%20 2000%20policy%20year>. Before the policy year 2000, the limit offered by P&I Clubs in the International Group was US $500 million for each occurrence for each vessel in the case of oil pollution; for charterers' insurance, it was US $300 million. P&I Clubs also made available to shipowners another US $200 million in oil pollution coverage from the commercial market. This limitation was a corresponding reflection of the market capacity for reinsurance. See *Hill/Robertson/Hazelwood*, Introduction to P&I (1996) 50. *Hazelwood* (Note 31 supra) 227; and *Gauci*, Oil Pollution at Sea – Civil Liability and Compensation for Damage (1997) 220.

35 For tanker-oil pollution liability, the Fund Convention provides a second-tire compensation fund.

4. Insurance and the question for adequate compensation

As previously stated, parties other than the registered owner, such as the bareboat charterer, manager, and operator, are not required to take out insurance. Therefore, although it would be possible for victims to claim against all other liable persons, the registered owner's insurance is likely to be the most frequently claimed compensation source, since it is the only guaranteed source of compensation. At the same time, the Bunkers Convention provides that victims are entitled to claim directly against the insurer.[36] This underscores the importance of the registered owner taking out insurance.

The Bunkers Convention aims to provide adequate compensation to victims suffering oil pollution damage. In light of this goal, one question is whether a supplementary compensation source is required in cases where the registered owner's insurance is inadequate. However, this kind of supplementary compensation source was not incorporated into the Bunkers Convention upon its advent, or at a later stage. Fortunately, to date no disastrous bunker-oil spill incident has been recorded. Any precautionary measures in this respect, including adequate compensation, are nevertheless needed.[37]

5. Insurance certificate and its validity

The Bunkers Convention requires that insurance or other financial security institutions be certified. One concern arising from this stipulation was that there would be huge administrative burden involved in issuing and inspecting insurance certificates due to the large number of different ships involved. Different proposals were advanced to alleviate this bureaucratic burden. The final provision included in the Bunkers Convention requires only that ships registered in a State Party having a gross tonnage greater than 1 000 maintain insurance. The result of settling a relatively high insurance threshold is that the number of ships required to take out insurance could be reduced and the administrative burden involved in issuing and inspecting insurance certificates thereby abated.

The issued certificate shall attest that insurance or other financial security is in force for the purpose of the Bunkers Convention. Article 7(2)(f) of the Bunkers Convention specifically provides that the "...period of validity of the certificate

---

[36] The Bunkers Convention, Article 7(10): "Any claim for compensation for pollution damage may be brought directly against the insurer or other person providing financial security for the registered owner's liability for pollution damage...".

[37] Causality experience in the tanker oil spill incidents may provide some estimation. The most recent *Prestige* disaster is a good example. The facts of the case can be summarised as follows: on 19 November 2002, the 26-year-old, single-hulled oil tanker, *Prestige*, sank and took 50,000 tons of its 77,000 tons of heavy fuel oil down to the bottom of the North Atlantic. It was indicated that the pollution victims of the *Prestige* disaster may never receive full compensation for the losses they have suffered. It can be imagined that if a large amount of fuel oil spills out from a cargo ship such as from a big container ship, the compensation amount envisaged in the Bunkers Convention will be inadequate. Therefore, some precautionary measures need to be considered.

which shall not be longer than the period of validity of the insurance or other security." More essentially, once liability has been established, the certificate must be valid to ensure the accessibility of compensation to eligible claimants. It is important to ensure to claimants at the time and place of an incident that: (1) the certificate concerned was issued by the authorised organisation of any State Party to the Convention; (2) the certificate embodies a guarantee of sufficient financial strength; and (3) the certificate can be valid at the time and place of the incident.

III. Conclusion

As discussed, some issues have emerged with respect to the liability and insurance aspects of the Bunkers Convention. Nevertheless, the Bunkers Convention closes the final significant gap in the international regime for compensating victims of oil spills from ships.[38] There are four advantages to this Convention:

(1) Ships of different types, except oil tankers, are covered for the purpose of dealing with bunker-oil pollution liability.
(2) The shipowner is strictly liable for pollution damage. He is liable simply because of the fact that his ship had an oil spill and caused pollution damage.
(3) The registered owner of a ship meeting the required gross tonnage shall compulsorily take out insurance.
(4) The shipowner and the shipowner's liability insurer are entitled to pay compensation up to the required limitation; limited liability may to some extent ensure the prompt payment of compensation without litigation.

Strict liability facilitates the victim's claim against the shipowner; however, this rule is futile unless the insurer or a similar financial institution is willing and competent to insure the liability. Although the availability of adequate and prompt compensation for victims is the paramount objective of the Bunkers Convention, this Convention does not provide for a supplementary compensation source. The registered owner's insurance represents the only guaranteed source of compensation for pollution victims.

There are so far 11 Contracting States to the Bunkers Convention representing 14.97% of world tonnage.[39] Compared with the requirement of 18 States to be for the ratification of the Bunkers Convention, there is still a long way to go.[40] Based on the need for a set of uniform rules to address liability and compensation for

[38] The information is available at <http://www.imo.org/Newsroom/contents.asp?topic_id=67&doc_id=457>. The former IMO Secretary-General, Mr. William A. O'Neil, pointed out: "The adoption of a bunkers convention completes the task initiated by the Legal Committee when it was established by IMO more than 30 years ago – namely, the adoption of a comprehensive set of unified international rules governing the award of prompt and effective compensation to all victims of ship-sourced pollution".

[39] See <http://www.imo.org/Conventions/mainframe.asp?topic_id=247> (visited on 5 May 2005). These 11 Contracting States include: Cyprus, Greece, Jamaica, Latvia, Luxembourg, Samoa, Singapore, Slovenia, Spain, and Tonga, the United Kingdom.

[40] The Bunkers Convention, Art.14(1).

victims of bunker oil pollution, it is hoped that the Bunkers Convention will enter into force, and that it will address the problems it is intended to solve.

Closing Remarks

Ulrich Magnus

Dear ladies and gentlemen, dear colleagues!

Our conference has now come to an end, and it is my task and privilege to say a few concluding words. I would like to start with the necessary thanks.

First, I would like to thank our speakers for their presentations, and all the participants in our discussions. A special thank you goes out to those responsible for the effective organisation of this Conference. I would like to mention Dr. Wurmnest in this respect. Finally, I would like to thank the representatives of this Tribunal for hosting us so well.

I would like to add a few observations. This Conference was a good reflection of the work done at the Research School for Maritime Affairs. Four characteristics of this work are evident:

First, interdisciplinary work is hard, and requires time and patience if different scholars studying natural sciences, economics, law are to understand each other and cooperate effectively. However, all sides profit from such efforts. The Conference and the Research School have both made good progress in improving interdisciplinary cooperation.

Interdisciplinary work is particularly indispensable in the field of maritime affairs. It is important to understand causal relationships between emissions, waste disposal at sea, and other such activities, and their consequences. One striking example of the need for this cooperation is the impressive number of conventions on pollution damage caused by ships in existence, despite the fact that 80% of pollution damage is caused by land-based activities which are not covered by any convention. Prof. Sündermann's report as well as Ms. Ilyina's and Ms. Röckmann's presentations made it clear that our knowledge about causal links remains limited, even though science continues to expand the factual basis of knowledge. However, as expressed by Alan Simcock, where sufficiently plausible results are offered, and reasonable grounds for concern exist, action should be taken, be it legal, economic, or otherwise. When considering protection of the marine environment, less than full conviction with respect to causation must sometimes be sufficient to justify a response. Otherwise, a successful reaction may come too late. We should be also aware that by no means is an adequate response to marine pollution necessarily the full prohibition of a certain activity. Protective measures limited in time or space must also be taken into account. In this sense, the precautionary principle, mentioned by many of our speakers, should be applied.

A second feature shared by the Conference and the Research School is their internationality. Maritime affairs are international by nature, and this Conference and the Research School pay tribute to this fact. The opportunity for people from different nations to discuss maritime problems, exchange their views, and find compromises, is indispensable. International cooperation between nations or nationals from different countries resembles interdisciplinary cooperation, in the sense that neither is always easy. However, international cooperation is rewarding, and the only real alternative. There is almost no local activity that does not affect distant regions via the medium of water. The example of Canadian plastic waste that drifts to the Norwegian shore is only one of many situations clearly demonstrating the globalised effects of environmentally relevant activities. Global implications require global rules. With respect to shipping, it is more or less understood that worldwide rules are necessary. The preservation and protection of the areas where shipping and all other marine activities occur also necessitates global standards.

The third feature that must be mentioned is the cooperation between different institutions. This Conference and the Research School exemplify such cooperation, which is also characteristic of maritime affairs. The fact we are here in this impressive building is due to the Research School's good contacts with the International Tribunal for the Law of the Sea. The Research School itself owes its existence to the intensive cooperation between the two Hamburg Max Planck Institutes and the University of Hamburg. With respect to the cooperation between international organisations in the field of protection of the marine environment, this conference has witnessed the expression of different views, in particular concerning activities of, and cooperation with the EU. Cooperation can probably always be improved. Rivalry between institutions, even where they compete to a certain extent, should not dictate their activities. However, it is not a contradiction to appeal for global standards, and to have specialised regional agencies for the implementation and administration of these standards. The presentations by Prof. Krämer, Alan Simcock, and Ms. Brusendorf gave the impression that working method for protective measures are at least similar: first, research and monitoring, second, the fixing of standards, and finally, the implementation and enforcement of these standards. In the end, a global approach appears advisable. As proposed by Judge Mensah, all measures reasonably expected to effectively protect the marine environment against dangers of serious concern should be taken.

However, the characteristic features thus far mentioned would be incomplete if another, decisive feature remained unmentioned, namely, innovation. Both the Conference and the Research School aim to develop new ideas. Although it is dangerous to make this type of assessment at the end of a conference, it is my firm conviction that this Conference and the Research School are successful in this respect. All of the presentations provided food for thought and initiated lively discussions that will certainly not end when we leave this room.

When we realise that we are sitting in the courtroom of an International Tribunal, and if we were to regard the Conference discussions as pleadings in a lawsuit, it becomes rather easy to identify the plaintiff: the aim of protecting the marine environment. The defendants would be numerous, namely, those activities which pollute the sea in a way that transgresses the regenerative powers of the maritime environment. The judge would be the environment itself. This judge never forgets, and never forgives. After almost two days of hearings, the judgment must now to be delivered: "It is hereby ordered that a reasonable use of the sea be continued and secured also for future generations. International and interdisciplinary efforts must be intensified to avoid marine pollution. This is particularly true with respect to persistent, toxic, or otherwise hazardous substances of any kind, which impede a reasonable and sustainable use of the environment. The costs must be borne by the defendants."

If I were a shark, as Prof. Krämer invited us all to become, and could read Hölderlin, I would follow Hölderlin in a slightly modified form: Europeans, be strong in reflection, but even stronger in action. The Conference is closed.

About the International Max Planck Research School for Maritime Affairs at the University of Hamburg

The International Max Planck Research School for Maritime Affairs at the University of Hamburg was established by the Max Planck Society for the Advancement of Science, in cooperation with the Max Planck Institute for Comparative and International Private Law (Hamburg), the Max Planck Institute for Comparative Foreign Public Law and International Law (Heidelberg), the Max Planck Institute for Meteorology (Hamburg) and the University of Hamburg. The School's research is focused on the legal, economic, and geophysical aspects of the use, protection, and organization of the oceans. Its researchers work in the fields of law, economics, and natural sciences. The School provides extensive research capacities as well as its own teaching curriculum. Currently, the School has 15 Directors who determine the general work of the School, act as supervisors for dissertations, elect applicants for the School's PhD-grants, and are the editors of this book series:

Prof. Dr. Dr. h.c. Jürgen Basedow is Director of the Max Planck Institute for Foreign Private Law and Private International Law; *Prof. Dr. Peter Ehlers* is the Director of the German Federal Maritime and Hydrographic Agency; *Prof. Dr. Dr. h.c. Hartmut Graßl* is Director of the Max Planck Institute for Meteorology; *Prof. Dr. Hans-Joachim Koch* is Managing Director of the Seminar of Environmental Law of the Faculty of Law at the University of Hamburg; *Prof. Dr. Rainer Lagoni* is Managing Director of the Institute of Maritime Law and the Law of the Sea at the University of Hamburg; *PD Dr. Gerhard Lammel* is Senior Scientist at the Max Planck Institute for Meteorology; *Prof. Dr. Ulrich Magnus* is Managing Director of the Seminar of Foreign Law and Private International Law at the University of Hamburg; *Prof. Dr. Peter Mankowski* is Director of the Seminar of Foreign and Private International Law at the University of Hamburg; *Prof. Dr. Marian Paschke* is Director of the Institute of Maritime Law and the Law of the Sea at the University of Hamburg; *PD. Dr. Thomas Pohlmann* is Senior Scientist at the Centre for Marine and Climate Research and Member of the Institute of Oceanography at the University of Hamburg; *Dr. Uwe Schneider* is Assistant Professor at the Research Unit Sustainability and Global Change, Centre for Marine and Climate Research, Departments of Geosciences and Economics at the University of Hamburg; *Prof. Dr. Jürgen Sündermann* is Director at the Centre for Marine and Climate Research at the University of Hamburg; *Prof. Dr. Richard Tol* is Director of the Research Unit Sustainability and Global Change at the University of Hamburg; *Prof. Dr. Dr. h.c. Rüdiger Wolfrum* is Director at the Max Planck Institute for Comparative Foreign Public Law and International Law and a judge at the International Tribunal for the Law of the Sea; *Prof. Dr. Wilfried Zahel* is professor at the Centre for Marine and Climate Research, Departments of Geosciences and Economics at the University of Hamburg.

At present, *Prof. Dr. Dr. h.c. Jürgen Basedow* and *Prof. Dr. Ulrich Magnus* serve as speakers of the International Max Planck Research School for Maritime Affairs at the University of Hamburg.

Printed by Printforce, the Netherlands